P9-BYT-297

STEM CELL *Research*

STEM CELL *Research*

New Frontiers in Science and Ethics

edited by

Nancy E. Snow

University of Notre Dame Press
Notre Dame, Indiana

Copyright © 2003 by University of Notre Dame
Notre Dame, Indiana 46556
www.undpress.nd.edu
All Rights Reserved

Reprinted in 2004

Chapter 2, "Human Embryonic Stem Cell Research: Ethics in the Face of Uncertainty," is printed here by permission of the author and Georgetown University Press. A version of this chapter appears in *God and the Embryo: Religious Voices on Stem Cells and Cloning,* edited by Brent Waters and Ronald Cole-Turner (Georgetown University Press, 2003).

Manufactured in the United States of America

Library of Congress Cataloging-in-Publication Data
Stem cell research : new frontiers in science and ethics / edited by Nancy E. Snow.
p. ; cm.
Includes bibliographical references and index.
ISBN 0-268-01778-6 (paper : alk. paper)
1. Stem cells—Research—Moral and ethical aspects.
[DNLM: 1. Ethics, Medical. 2. Stem Cells. 3. Ethics, Research 4. Religion and Medicine. W 50 S818 2003] I. Snow, Nancy E.
QH588.S83 S74 2003
174.2'8—dc22

2003020077

∞ *This book is printed on acid-free paper.*

Contents

STEM CELL *Research*

Introduction

Stem Cell Research: New Frontiers in Science and Ethics

In October 2001, a unique conference was held in Milwaukee, Wisconsin. Cosponsored by the Archdiocese of Milwaukee, Marquette University, and the Wisconsin Catholic Conference, it brought together an internationally distinguished group of scientists and ethicists to discuss the scientific, ethical, and public policy implications of stem cell research. At that time, debates about stem cell research were well underway. Fearing that discussions about human embryonic stem cell research would become as polarized as the abortion debate, the conference sponsors sought to bring together presenters and panelists with a variety of perspectives on the issues. The aim was to generate more light than heat—to move the discussion on stem cell research forward. In achieving this aim, the conference was successful. It also produced the essays on the science, ethics, and public policy of stem cell research collected in this volume, which derive from panel contributions by David A. Prentice, John Langan, S.J., and Ronald M. Kline, and from presentations by Kevin FitzGerald, S.J., Ira B. Black, Karen Lebacqz,

Edward J. Furton, Lisa Sowle Cahill, Richard M. Doerflinger, and M. Therese Lysaught.

BACKGROUND OF THE STEM CELL RESEARCH CONTROVERSY

The stem cell research controversy was prompted by the almost simultaneous announcements in 1998 of the isolation of human embryonic stem cells by Dr. James Thomson and human embryonic germ cells by Dr. John Gearhart.[1] Stem cells derived from these sources are pluripotent, that is, capable of differentiating into virtually every human cell type.[2] Because of this remarkable plasticity, scientists believe that stem cell research will eventually yield therapies capable of treating stroke, diabetes, Alzheimer's, Parkinson's, and a host of other currently incurable diseases.[3] However, because human embryonic stem cells cannot be derived without destroying embryos, human embryonic stem cell research has provoked ethical and religious objections. Since human embryonic germ cells are derived from fetal tissue, the use of aborted fetuses as a source of germ cells also raises ethical questions. Consequently, President Clinton requested the National Bioethics Advisory Commission (NBAC) to study issues raised by stem cell research. NBAC held public hearings during the spring of 1999 and published its report in 1999 and 2000.[4] This comprehensive report covers scientific, legal, ethical, religious, and public policy perspectives on stem cell research and offers a series of policy recommendations. The policy of the Clinton administration regarding the federal funding and regulation of stem cell research was to permit federal funding for stem cell research using aborted fetal tissue, placentas and umbilical cord blood, and adult stem cell research.[5] It also allowed funding for research using stem cells that had been derived from frozen embryos created for the purpose of fertility treatment that were in excess of clinical need, but it prohibited research in which embryos were destroyed to obtain the cells. The Clinton administration also prohibited federal funding for the creation of embryos for research purposes through somatic cell nuclear transfer (SCNT); for research using stem cells derived from embryos created through SCNT; and for research involving animal-human hybrids.

Debates about the ethics of stem cell research continued. Critics challenged the Clinton policy. Some viewed as specious the distinction between research involving the derivation of stem cells through the destruction of embryos and that involving the use of previously derived embryonic stem cells.[6] On August 9, 2001, President Bush announced a compromise position on the federal funding of human embryonic stem cell research: federal monies will be available for research on seventy-eight currently existing stem cell lines.[7] Federal funds will also support adult stem cell research and research using stem cells derived from umbilical cord blood and placentas. Research involving the creation or new destruction of human embryos will not be eligible for federal financing. Federal funding for stem cell research using aborted fetal tissue is unaffected by Bush's decision and, consequently, this research is still eligible for federal funding.

BACKGROUND ON STEM CELL RESEARCH

What are stem cells? According to a report on the science of stem cell research issued by the National Institutes of Health (NIH) in the summer of 2001, "A stem cell is a special kind of cell that has a unique capacity to renew itself and to give rise to specialized cell types."[8] Stem cells can be derived from embryos and fetal tissue. Adult stem cells have also been identified. Embryonic stem cells are derived from a group of cells called the inner cell mass, which is part of the early (four– to five-day-old) embryo, known as a blastocyst. The blastocyst consists of about 200 to 250 cells.[9] Embryonic stem cells are not embryos. The embryos from which these cells are derived are destroyed in the process.

Another potential source of stem cells is embryos created through somatic cell nuclear transfer, or therapeutic cloning.[10] In this process, chromosomal material from a somatic cell—that is, any cell other than an egg or sperm cell—is transferred into an egg cell from which the nucleus, which contains the chromosomal material, has been removed. The fused cell and its immediate descendants are believed to be totipotent, that is, to have the full potential for developing into a complete animal. These totipotent cells will form a blastocyst, whose inner cell mass could be used to derive pluripotent stem cells.

Stem cells derived from fetal tissue are called embryonic germ cells. They are isolated from the primordial germ cells of the gonadal ridge of the five– to ten-week-old fetus. Later in development, the gonadal ridge becomes the testes or ovaries, and the primordial germ cells become sperm or eggs. Embryonic stem cells and embryonic germ cells are pluripotent, or capable of differentiating into virtually every cell type in the human body.[11] However, the two types of cells do not have identical properties.

Adult stem cells are multipotent, that is, capable of differentiating into some, but not all, cell types. Adult stem cells have been found in bone marrow, blood, the cornea and the retina of the eye, the brain, skeletal muscle, dental pulp, liver, skin, the lining of the gastrointestinal tract, and the pancreas.

Stem cells have research applications, for example, in the study of embryonic development and in cancer research.[12] Stem cell lines could also be used to streamline the process of drug testing. Specialized cells and tissues generated from stem cells could be used for stem cell–based therapies to treat diseases such as Parkinson's, Alzheimer's, spinal cord injury, stroke, burns, heart disease, diabetes, osteoarthritis, and rheumatoid arthritis. According to the NIH report, "To date, it is impossible to predict which stem cells—those derived from the embryo, the fetus, or the adult—or which methods for manipulating the cells, will best meet the needs of basic research and clinical applications."[13]

Why is stem cell research important and controversial? As noted, proponents believe that stem cell research promises enormous clinical benefits and will greatly enhance our scientific knowledge. They maintain that only embryos in excess of clinical need will be used for research purposes. These embryos are slated for destruction anyway. Why not allow their use for beneficial purposes? The destruction of embryos, which is required for the extraction of embryonic stem cells, is precisely the practice to which opponents object.[14] Opponents believe that embryos have rights to life. Consequently, destroying them is an act of murder. Additionally, some object to the use of aborted fetal tissue as a source of embryonic germ cells, maintaining that research use of this tissue is complicity in an evil act and could encourage abortion. Opponents point to adult stem cell research as a readily available and promising alternative. Further, ethicists and theologians have worried that therapies developed from stem cell research will not be justly distributed.

This volume is divided into two parts. Part I, "Scientific and Public Policy Perspectives," consists of the essays by Prentice, FitzGerald, Langan, Kline, and Black and Woodbury. Part II, "Ethical Issues in Stem Cell Research," consists of the essays by Lebacqz, Furton, Cahill, Doerflinger, and Lysaught. A brief interpretative essay, intended to guide the reader through the main arguments offered by contributors, introduces each part.

The complexity of the issues raised by stem cell research is daunting. Easy answers will not readily be found. Contributions to this volume move the discussion forward through the depth and breadth of their reflections on the debate. They remind us of the interconnectedness of science, ethics, religion, law, and public policy. They remind us, too, that the stem cell research controversy is related to other pressing issues, such as infertility medicine and the treatment of the early embryo, therapeutic and reproductive cloning, and genetic engineering, especially germline interventions. Greater awareness and public discussion are needed to advance these debates. This volume is one contribution to this enterprise.

Nancy E. Snow
Marquette University

NOTES

1. See James A. Thomson et al., "Embryonic Stem Cell Lines Derived from Human Blastocysts," *Science* 282 (1998): 1145–47; and M. J. Shamblott et al., "Derivation of Pluripotent Stem Cells from Cultured Human Primordial Germ Cell," *Proceedings of the National Academy of Sciences* 95 (1998): 13726–31.

2. See NIH (National Institutes of Health), *Stem Cells: Scientific Progress and Future Research Directions* (June 2001), ES-2. Unless otherwise indicated, the discussion of the science of stem cell research in this introduction is summarized from the "Executive Summary" of this report.

3. See ibid., 4–6; and NIH, "Stem Cells: A Primer," www.nih.gov/news/stemcell/primer.htm, May 2000, 3.

4. See National Bioethics Advisory Commission, *Ethical Issues in Human Stem Cell Research,* 3 vols. (Rockville, Md.: National Bioethics Advisory Commission, 1999–2000).

5. For a complete statement of the Clinton administration's guidelines, see NIH, "NIH Fact Sheet on Human Pluripotent Stem Cell Guidelines," NIH, www.nih.gov/news/stemcell/factsheet.htm, 23 August 2000, 1–3.

6. See NIH, "National Institutes of Health Guidelines for Research Using Human Pluripotent Stem Cells," www.nih.gov/news/stemcell/stemcellguidelines.htm, 24 August 2000, 1.

7. See "Excerpts from Bush Address on U.S. Financing of Embryonic Stem Cell Research," *New York Times,* 10 August 2001, A16; Katherine Q. Seelye, "Bush Gives His Backing for Limited Research on Existing Stem Cells," *New York Times,* 10 August 2001, A1, A16; George W. Bush, "Stem Cell Science and the Preservation of Life," *New York Times,* 12 August 2001, sec. 4, p. 13. At the time the policy was announced, the number of stem cell lines meeting Bush's criteria was sixty. At the time of this writing, the number is seventy-eight. For a complete statement of the Bush administration guidelines, see "NIH Human Embryonic Stem Cell Registry," U.S. Department of Health and Human Services, National Institutes of Health, escr.nih.gov, 1.

8. NIH, *Stem Cells: Scientific Progress and Future Research Directions,* ES-1.

9. See Robert Pear, "Health Institutes Study Praises the Potential of Stem Cells in Medicine," *New York Times,* 27 June 2001, A12; also Pear, "G.O.P. Leaders in House Fight Stem-Cell Aid," *New York Times,* 3 July 2001, A10.

10. See NIH, "Stem Cells: A Primer," 2–3.

11. For other properties of embryonic stem cells, see NBAC (National Bioethics Advisory Commission), *Ethical Issues in Human Stem Cell Research,* vol. 3, *Religious Perspectives* (Rockville, Md.: National Bioethics Advisory Commission, June 2000), Testimony of Laurie Zoloth, Ph.D., J-5.

12. See NIH, *Stem Cells: Scientific Progress and Future Research Directions,* ES-5; and NIH, "Stem Cells: A Primer," 3–4.

13. NIH, *Stem Cells: Scientific Progress and Future Research Directions,* ES-10.

14. See, for example, "Roman Catholic Views on Research Involving Human Embryonic Stem Cells," testimony of Margaret A. Farley, Ph.D., in NBAC, *Ethical Issues in Human Stem Cell Research,* vol 3, *Religious Perspectives,* D-3–D-5.

PART I

Scientific and Public Policy Perspectives

Part I presents a variety of scientific and public policy perspectives on stem cell research. In "The Present and Future of Stem Cell Research: Scientific, Ethical, and Public Policy Perspectives," David A. Prentice identifies the fundamental questions at stake in the stem cell research debate as those also raised by cloning, genetic engineering—especially germline interventions—and the creation of animal-human hybrids: What does it mean to be human? To whom will—or should—we assign value? Who will decide? And who will benefit from these choices?

Prentice suggests that the primary question in the stem cell research debate is the moral status of the human embryo. Whether one views the embryo as realized or potential human life, the use of human embryos for research purposes or as a source of potential clinical materials raises questions of commodification and risks diminishing respect for all human life. Prentice also cautions against conflating scientific claims with scientific facts. He is skeptical about arguments made in favor of human embryonic stem cell research. To date, he argues, scientific claims made on behalf of the potential of human embryonic stem cell research remain unsubstantiated, unlike claims made on behalf of adult stem cell research. Moreover, serious scientific questions about human embryonic stem cells remain unanswered. For example, embryonic stem cells are difficult to establish and maintain in culture. They are genomically unstable; that is, they exhibit variable gene expression even under controlled conditions. No clinical treatments using embryonic stem cells exist, and potential treatments using these cells or tissues grown from them risk tumor formation and immune rejection by the patient. Further, Prentice points out that the destruction of frozen embryos is neither inevitable nor does it occur at the rate some have claimed. These and other reasons cause him to wonder: why the rush to advance human embryonic stem cell research?

How much do we really know about stem cell research? Kevin T. FitzGerald, S.J., argues that we know less than we think. He explores multiple uncertainties that attend the stem cell research debate in "Human Embryonic Stem Cell Research: Ethics in the Face of Uncertainty." Advances in genetics and regenerative medicine, he contends—especially brain research—force us to examine questions of the uniqueness or significance of human nature. Questions about human nature, about who or what counts as human, are raised by the specter of human embryonic stem cell research. Debates about the value of this research are also plagued by ambiguities in the scientific and ethical conceptualizations of the term "embryo." Beyond that, we must ask whether the potential benefits of stem cell research justify the harm to human embryos that such research requires.

FitzGerald examines two arguments often made in support of human embryo research: that there is a need for such research and that large numbers will benefit from it. Neither argument, he claims, is compelling, for each is beset by uncertainties. First, it is not clear that there is a pressing need for human embryonic stem cell research, since illnesses that might be treated by such research could well be addressed more quickly through other avenues. Second, there is no direct correlation between advances in human embryonic stem cell research and medical benefits for those who need them. This is evident from the realities of health care systems in the United States and around the world. We must face the hard reality that, even if human embryonic stem cell research is the first to yield medical treatments, many, if not most, of those who need them will not receive them.

In the face of the pervasive uncertainties found in all aspects of stem cell research, how should society proceed? FitzGerald proposes that less controversial alternatives to human embryonic stem cell research should receive government funding, and that expert recommendations for improving worldwide health care should receive heightened attention and action.

In "Stem Cell Research and Religious Freedom," John Langan, S.J., considers the question of how stem cell research should be treated in a pluralistic society. Specifically, does the Catholic Church's commitment to religious freedom have implications for stem cell research within the domain of public policy? Langan argues that it does.

He considers questions of stem cell research within the framework of the natural law tradition. Drawing on the natural law theory of St. Thomas Aquinas, Langan identifies three levels of natural law principles: highly general precepts such as "good is to be done, and evil avoided"; specific precepts forbidding certain kinds of actions, such as the prohibition against killing; and further conclusions and specifications of these precepts. Principles of the first type, according to Aquinas, are known to all who understand the meaning of key terms. Failure to acknowledge principles of the second type, Aquinas thinks, creates social division and threatens the stability of the community. Principles of the third type are, in Aquinas's view, known only to the wise or the learned.

Though Catholics and conservative Christians often treat advocates of human embryonic stem cell research as rejecting principles of the second type, Langan argues that there is reason to treat the stem cell research debate as implicating principles of the third type. Many who disagree with Catholic and conservative approaches to human embryonic stem cell research do so on the basis of morally weighty reasons involving requirements of conscience and setting directions for their own exercise of religious freedom.

Given that widespread societal consensus on the ethics of human embryonic stem cell research is unlikely, how should Catholics approach this debate? Langan suggests and explains three guidelines. First, Catholic institutions should not participate in human embryonic stem cell research. This guideline applies to the present scenario, in which the moral acceptability of human embryonic stem cell research to the Catholic tradition has not been shown. Second, Catholics should be prepared to tolerate this research, while critically questioning it. Catholics should take seriously differing perspectives on stem cell research and treat with respect and charity those who hold them. Third, government should not fund this research out of respect for the consciences and the religious freedom of those who oppose it. If proponents of human embryonic stem cell research cannot generate widespread consensus in its favor, and if large groups continue to voice conscientious objections to it, proponents should be willing to acknowledge the good of showing respect for the consciences of opponents. This good, which is essential for a pluralistic democracy, is often obscured from view in the heat of controversy.

Alternatives to human embryonic stem cell research are being studied. In "Umbilical Cord Blood, Stem Cells, and Bone Marrow Transplantation," Ronald M. Kline, M.D., indicates that umbilical cord blood could provide a virtually limitless source of pluripotent stem cells and thereby furnish an alternative to ethically controversial research on human embryonic stem cells. Using umbilical cord blood as a source of hematopoietic stem cells promises to improve blood and marrow transplantation (BMT). Kline sketches the history of bone marrow transplantation. One of BMT's greatest limitations, he asserts, is the scarcity of suitable donors. These problems were partially addressed in 1986 with the establishment of the National Marrow Donor Program to facilitate unrelated donor transplants in the United States. However, the NMDP is successful in identifying a donor only 75 percent of the time; the percentage is even less in minority patients. Umbilical cord blood became available as a source of hematopoietic stem cells during the last decade, and it has, Kline thinks, the potential to alleviate the shortage of donors that plagues BMT.

Kline reviews the clinical experience of umbilical cord blood transplantation, including advantages and disadvantages. Because cord blood harvested from fetuses is an untested source of hematopoietic stem cells, among the greatest potential disadvantages is the inadvertent transmission of genetic diseases. He argues that congenital abnormalities could be detected by the active clinical follow-up of cord blood donors six to twelve months after birth. This policy, however, raises privacy concerns. In addition to privacy, other ethical issues raised by unrelated cord blood donation include the questions of who owns the cord blood and how this limited resource should be allocated. The existence of "for profit" umbilical cord banks raises unique ethical issues, most notably the ethics of marketing an expensive service to new parents who pay to have cord blood stored when the probability that they will need the blood is low. Despite these concerns, Kline is optimistic about the potential of pluripotent stem cells derived from umbilical cord blood to provide a viable alternative to human embryonic stem cells.

In "Stem Cell Plasticity: Adult Bone Marrow Stromal Cells Differentiate into Neurons," scientists Ira B. Black and Dale Woodbury describe experiments in which they have induced a particular type of adult stem cell, called bone marrow stromal cells (BMSCs), to differentiate into neurons. Experiments on rat and human BMSCs suggest that adult BMSCs can differentiate into neurons both in vitro and in vivo. According to Black and Woodbury, these findings change our conception of normal brain function and raise possibilities for finding treatments for currently hopeless brain disease, such as Parkinson's, Alzheimer's, and Lou Gehrig's diseases, as well as stroke and spinal cord injury.

They caution, however, that these findings are still rudimentary. Many scientific, clinical, and ethical questions remain unanswered. For example, how can immunorejection of foreign cells and tissues be prevented and the use of toxic immunosuppressants be avoided? How can enough cells be generated for each patient for the treatment of chronic illnesses? Can safe, accessible sources of cells for long-term therapy be found? Can cells be generated in ways that minimize tumor formation? Which stem cell populations are the most advantageous for clinical use? How do embryonic, fetal, and adult stem cell populations compare?

Comparative studies using embryonic, fetal, and adult cells are just beginning to address critical issues. Unless these concerns are dealt with, it will be difficult, they warn, to formulate therapeutic approaches that maximize effectiveness and minimize side effects and toxicity. Nevertheless, we are, they claim, now presented with an exceptional opportunity to reformulate mechanisms of development and protocols for treatment.

CHAPTER
1

The Present and Future of Stem Cell Research

Scientific, Ethical, and Public Policy Perspectives

DAVID A. PRENTICE

Many considerations come into play when considering the question of using human embryos as a source of stem cells for potential treatments. The moral status of the human embryo is the primary consideration in the debate. Scientifically, genetically, we are human beings even at the one-cell stage. What follows is a developmental continuum through various stages, but with no readily definable breakpoint in our developmental program. Moreover, the question of the presence of the human soul, or human personhood, is outside of the scope of science; this is instead a theological and philosophical definition. Nonetheless, whether one views a human embryo as realized or potential human life, use of human embryos for research or as a source of potential clinical materials raises the specter of commodification of human life, lessening the value of all human life.

Another aspect in the debate regarding destructive human embryo research is the distinction between scientific claims and scientific facts. Too often a false choice has been put forth—

that embryos must be destroyed for potential medical benefits, or else patients will suffer. This ignores the alternatives to embryonic stem cells, for example, adult stem cells, which also promise significant medical benefits. On examining the promises, the premises, and the actual published data regarding embryonic stem cells, there is little published evidence on which to base the potential for significant medical benefits. Indeed, previous predictions of medical cures, such as using fetal tissue transplantation, have been made by many of those now promising similar cures with human embryonic stem cells. Yet after ten years of legalized and funded fetal tissue research, the cures promised with fetal tissue remain unrealized, and in fact the only published controlled study shows not only no real benefit to patients, but a worsening of the condition of a significant portion of patients receiving fetal tissue transplants.[1]

Other aspects of human embryo and embryonic stem cell use include the economics of the research and development, the intellectual property rights, possible regulatory oversight of the research, and public access to any benefits realized. A key consideration is the question of scientific freedom to pursue any research path versus scientific stewardship of discoveries and their potential impact on society.

In September 1999 the Clinton National Bioethics Advisory Commission issued its report, "Ethical Issues in Human Stem Cell Research." Their recommendation was that use of human embryos for stem cell research should proceed; but they also noted:

> In our judgment, the derivation of stem cells from embryos remaining following infertility treatments is justifiable only if no less morally problematic alternatives are available for advancing the research. . . . The claim that there are alternatives to using stem cells derived from embryos is not, at the present time, supported scientifically. We recognize, however, that this is a matter that must be revisited continually as the demonstration of science advances.[2]

At that point in time, September 1999, there was little evidence that viable alternatives to embryonic stem cells existed. But has the science advanced, and is there now evidence for "less morally problematic alternatives"?

Let us first consider the basic science of stem cells. A stem cell has two chief characteristics: (1) it can continue to proliferate, maintaining a pool of growing stem cells; and (2) upon the correct signal, it can differentiate into one or more specific cell and tissue types. If we look through the human developmental continuum, there are numerous possible sources of stem cells. The source receiving most of the attention is the early human embryo, approximately five to seven days after conception. The inner cell mass of the early embryos is composed of embryonic stem cells, which have the capacity during normal development to form all tissues of the adult body. Other sources include embryonic germ cells, derived from the germ cell progenitors of later fetuses (six to nine weeks after conception), fetal tissue stem cells, umbilical cord blood, placenta, and adult stem cells. The term "adult" stem cell is a bit of a misnomer, since these tissue stem cells reside in most, if not all, of the body tissues even at birth.

Are adult stem cells and other nonembryonic sources viable alternatives to embryonic stem cells? The published scientific literature of the last few years indicates that these adult stem cells are indeed able to accomplish the transformation into various tissue types as is promised for embryonic stem cells. Several published papers now document that adult stem cells can generate most, if not all, of the adult body tissues, starting with either bone marrow or neural stem cells.[3] The published literature now indicates that most tissues either contain their own adult stem cell, or can be formed from an adult stem cell in another tissue.[4] Adult stem cells are also currently being used successfully to treat patients. Some of these clinical treatments are outgrowths of work with bone marrow transplants. Now, bone marrow stem cells are used in treatments for various cancers; autoimmune diseases such as lupus, multiple sclerosis, and arthritis; anemias such as sickle-cell anemia; and immunodeficiencies.[5] Beyond this, adult stem cells are also being used to grow new cartilage, bone, and corneas for patients; to repair cardiac tissue after a heart attack; and to grow new skin.[6] An added advantage of using the patient's own adult stem cells is the avoidance of possible transplant rejection.

The scientific literature is also burgeoning with reports showing the efficacy of adult stem cells to treat animal models of disease, including heart damage, stroke damage, diabetes, and muscular dystrophy. Thus, the majority of the scientific literature supports the contention that adult stem cells are a less morally problematic alternative.[7]

What of the literature regarding embryonic stem cells? Despite twenty years of experiments with mouse embryonic stem cells, and despite the fervor that has arisen regarding human embryonic stem cells, the claims for embryonic stem cells and their advantages over adult stem cells are unsubstantiated. There are no current clinical treatments using embryonic stem cells; there are few and modest advances with embryonic stem cells in animal models of disease; and there is even difficulty obtaining pure cultures of cells in the laboratory dish. The embryonic stem cells themselves are difficult to establish and maintain in culture. Proponents of embryonic stem cell use admit that potential treatments run the risk both of immune rejection by the patient and of tumor formation.[8] And a report published in the summer of 2001 noted that embryonic stem cells are genomically unstable, exhibiting variable gene expression even in the controlled conditions of the laboratory.[9]

It is difficult then to understand why there is a continued push for the destruction of human embryos to obtain embryonic stem cells. The situation is compounded when supposed scientific experts weigh in with their opinions, which do not seem to be based on factual evidence but, rather, simply a fervor for unhindered scientific inquiry. As a case in point, eighty Nobel laureates sent a letter in the spring of 2001 to President Bush encouraging federal funding of human embryonic stem cell research and citing numerous advances in stem cell science. However, most of the advances cited were using adult stem cells, and the vast majority of those signing the letter had no background related to the science discussed. An additional example involves the report on stem cells published by the National Institutes of Health in June 2001.[10] The report appears hastily compiled and is internally contradictory; just one example is the statement on page ES-6 that "no adult stem cell has been shown to be capable of developing into cells from all three embryonic germ layers." On the very next page (ES-7), the report states "blood and bone marrow (unpurified hematopoietic) stem cells differentiate into the three major types of brain cells (neurons, oligodendrocytes, and astrocytes) [ectoderm, one of the three embryonic germ layers], skeletal muscle cells, cardiac muscle cells [both derived from mesoderm, a second embryonic germ layer], and liver cells [endoderm, the third embryonic germ layer]." It would appear that these are not unbiased scientific arbiters in the debate. Another aspect in the embryonic stem cell debate has been the fate of existing embryos, the so-called "excess" embryos produced during in vitro fertilization treatments. While the phrase "they're going to be discarded anyway, let's get some good out of them" has been repeated often, there is actually no evidence of large-scale destruction of human embryos, making this a sort of "urban myth." While some embryos are undoubtedly discarded, the destruction is not nearly in the amounts proponents would have us believe. And, in fact, the destruction of frozen embryos is not an inevitability. Embryos can survive for very long periods, perhaps indefinitely, in the frozen state; they do not "go bad" in the freezer as might a steak. Published data show that embryos can be thawed after being frozen at least seven or eight years, implanted, and achieve a pregnancy and live birth.[11] Further, the disposition of embryos currently frozen (estimated at 100,000 in the United States) has a new option—embryo adoption. While a relatively new idea (the first frozen embryo adopted was in 1998; Hannah was born December 31, 1998), the option of embryo adoption is becoming more widely known and available.[12]

The real root of the debate is this: What does it mean to be human? As we look at different forms or stages of human life, to whom will we choose to assign value? Who will decide, and who will benefit, from these value choices?

These same basic questions regarding human life and human value return when we examine the issue of human cloning. In this case, the human embryo is specifically created to be virtually genetically identical to an existing or previously existing individual. The cloning (somatic cell nuclear transfer) technique creates a one-cell embryo that is then allowed to develop to an early stage, the blastocyst; this is the same stage at which embryonic stem cells are derived via the destruction of the embryo. The same questions arise not only as to the science and potential medical benefits involved in this process, but again regarding the basic value and expectation placed on a human life.

These debates will continue on into the future, with legislative and legal challenges, and, of course, with mounting scientific evidence. At present the mounting evidence does not favor use of human embryos for research purposes, even if a utilitarian argument were the only one espoused. Beyond stem cells, the specific creation of fertilized embryos for research use, cloning of human embryos, genetic engineering (especially germline genetic engineering), and chimerism (animal-human hybrids) will add new topics and nuances to the debates. However, the basic question will still be the same: What does it mean to be human, and to whom will we choose to assign value?

NOTES

1. Curt R. Freed et al., "Transplantation of Embryonic Dopamine Neurons for Severe Parkinson's Disease," *New England Journal of Medicine* 344 (8 March 2001): 710; Gina Kolata, "Parkinson's Research Is Set Back by Failure of Fetal Cell Implants," *New York Times,* 8 March 2001, A1.

2. National Bioethics Advisory Commission, *Ethical Issues in Human Stem Cell Research,* 3 vols. (Rockville, Md.: National Bioethics Advisory Commission, 1999), 1:53.

3. D.L. Clarke et al., "Generalized Potential of Adult Neural Stem Cells," *Science* 288 (2 June 2000): 1660; D.S. Krause et al., "Multi-Organ, Multi-Lineage Engraftment by a Single Bone Marrow–Derived Stem Cell," *Cell* 105 (4 May 2001): 369.

4. H.M. Blau et al., "The Evolving Concept of a Stem Cell: Entity or Function?" *Cell* 105 (2001): 829.

5. See, for example, I.J. Dunkel, "High-Dose Chemotherapy with Autologous Stem Cell Rescue for Malignant Brain Tumors," *Cancer Investigation* 18 (2000): 492; O.N. Koc et al., "Rapid Hematopoietic Recovery after Coinfusion of Autologous-Blood Stem Cells and Culture-Expanded Marrow Mesenchymal Stem Cells in Advanced Breast Cancer Patients Receiving High-Dose Chemotherapy," *Journal of Clinical Oncology* 18 (2000): 307; P. Nieboer et al., "Long-Term Haematological Recovery following High-Dose Chemotherapy with Autologous Bone Marrow Transplantation or Peripheral Stem Cell Transplantation in Patients with Solid Tumours," *Bone Marrow Transplantation* 27 (2001): 959; R.K. Burt, and A.E. Traynor, "Hematopoietic Stem Cell Transplantation: A New Therapy for Autoimmune Disease," *Stem Cells* 17 (1999): 366; G.L. Mancardi et al., "Autologous Hematopoietic Stem Cell Transplantation Suppresses Gd-Enhanced MRI Activity in MS," *Neurology* 57 (2001): 62; N.M. Wulffraat et al., "Prolonged Remission without Treatment after Autologous Stem Cell Transplantation for Refractory Childhood Systemic Lupus Erythematosus," *Arthritis & Rheumatism* 44 (2001): 728; P. Amrolia et al., "Nonmyeloablative Stem Cell Transplantation for Congenital Immunodeficiencies," *Blood* 96 (2000): 1239.

6. E.M. Horwitz et al., "Transplantability and Therapeutic Effects of Bone Marrow–Derived Mesenchymal Cells in Children with Osteogenesis Imperfecta," *Nature Medicine* 5 (1999): 309; R.J.-F. Tsai et al., "Reconstruction of Damaged Corneas by Transplantation of Autologous Limbal Epithelial Cells," *New England Journal of Medicine* 343 (2000): 86; P. Menasché et al., "Myoblast Transplantation for Cardiac Repair," *Lancet* 357 (2001): 279; B.E. Strauer et al., "Myocardial Regeneration after Intracoronary Transplantation of Human Autologous Stem Cells following Acute Myocardial Infarction," *Dtsch. Med. Wochenschr.* 126 (2001): 932; A. Limat et al., "Successful Treatment of Chronic Leg Ulcers with Epidermal Equivalents Generated from Cultured Autologous Outer Root Sheath Cells," *Journal of Investigative Dermatology* 107 (1996): 128.

7. For a more complete list of references, see the Do No Harm website (www.stemcellresearch.org).

8. Jon S. Odorico et al., "Multilineage Differentiation from Human Embryonic Stem Cell Lines," *Stem Cells* 19 (2001): 193.

9. D. Humpherys et al., "Epigenetic Instability in ES Cells and Cloned Mice," *Science* 293 (2001): 95.

10. National Institutes of Health, *Stem Cells: Scientific Progress and Future Research Directions* (June 2001), Washington, D.C.

11. Snunit Ben-Ozer and Michael Vermesh, "Full Term Delivery following Cryopreservation of Human Embryos for 7.5 Years," *Human Reproduction* 14 (June 1999): 1650.

12. See Snowflakes Embryo Adoption Program (www.snowflakes.org).

CHAPTER 2

Human Embryonic Stem Cell Research

Ethics in the Face of Uncertainty

KEVIN T. FITZGERALD, S.J.

Too often the opposing positions in the stem cell and cloning debate are presented in terms of the obviousness of their assertions made. Considering the complex nature of these controversial issues challenging our society, the reality is much less clear and certain. Therefore, the question addressed in this essay is: how might we best respond to the challenge of human embryonic stem cell research in the face of the uncertainties that pervade this issue?

Uncertainty is present in all aspects of this issue: scientific, medical, moral, religious, and political. This essay begins with the areas of uncertainty that are perhaps most surprising and, hence, most vexing for those engaged in this public debate—the scientific and the medical.

In order to more fully appreciate the scientific and medical uncertainties within stem cell research, it is helpful to put the science of the stem cell debate within the larger context of current advances in molecular and cellular research. Stem cells are only one part of the rapidly expanding arena of molecular biology research. This arena includes such topics as genetic therapies, genomics, pharmacology, proteomics, and various types of cellular and tissue research.[1] All of these research trajectories offer tremendous potential for advancing our scientific knowledge as well as the possibility of leading to new and exciting medical therapies and products. A couple of examples may help give a sense of the scope of these possibilities.

Much has been written both in academic journals and in the popular press about the promise of human gene therapy. Until recently there has been little concrete evidence of the fulfillment of that promise, and, instead, some tragic and troubling research tragedy has occurred.[2] However, the latest results of some clinical trials employing gene therapies to treat immune system disorders indicate that the promise might be at least partially satisfied.[3]

In these clinical experiments, the researchers added a functioning gene to the cells of individuals who were diseased because of a genetic defect. The drawbacks to this approach include the problem of not being able to control where the new gene incorporates in a cell's DNA and the problem of the mutated gene remaining in the cells. In the near future, researchers hope to address these problems by directly replacing or repairing the diseased genes.[4]

If the disease to be treated results from flaws in a region of DNA much larger than a single nucleotide or even a single gene, then researchers may try to employ artificial chromosomes to address the situation. In addition to this larger genetic carrying capacity, human artificial chromosomes would also have the advantages of maintaining a more stable number of the copies of a gene within a cell, along with better control of long-term gene expression.[5] Using these genetic technologies to target both small and large genetic mutations, the physicians of the future may have much greater success in treating the genetic causes of many diseases.

However, if these genetic technologies can be used successfully to treat disease, might not they also be used to change the genetic constitution of a human being in order to alter that individual's physical or behavioral characteristics? Considering the fact that human beings share over 98 percent of their genetic sequence with chimpanzees, the question arises: Would changing an individual's genetic constitution to include DNA sequences that are not known to have been present before in a human being change the nature of that individual? Is there an amount or kind of alteration that would result in that individual no longer being human?

These examples from human genetic engineering have been used to show that the ethical challenges generated by cutting-edge biotechnologies are very much the same as those raised by human stem cell research. An example from stem cell research will help to demonstrate this point.

In an experiment designed to investigate the emergence of reservoirs of neural stem cells in the developing fetal brain, Evan Snyder and Curt Freed directed research whereby cells from a neural stem cell line derived from a human fetal cadaver were implanted into the developing brain of a fetal bonnet monkey at approximately twelve to thirteen weeks of gestation. After sacrificing the fetal monkey four weeks later, Freed and Snyder found that the human neural stem cells had migrated and incorporated into the fetal monkey brain.[6]

Though it was not the stated purpose of the experiment, these results pose serious questions about the uniqueness or significance of human nature. If human beings are considerably interchangeable with other animals on a cellular and/or genetic level, then how might that reality affect our concepts of our selves? If we now have the ability to interchange genes, or cells, or even tissues and organs with other animals, then at what point does an addition of nonhuman cells, for example, make a human being something or someone else? Already researchers add human DNA and human cells to animals. In light of these realities, one could try to frame the question of human nature in terms of percentages of human DNA, cells, or genes expressed in a given animal. This approach, I would argue, is not likely to succeed because such quantification cannot encompass the complexity and richness found in our concepts of human nature.

The above experiment with human neural stem cells in fetal monkey brains is particularly relevant to these questions because arguments are often presented that focus on the human brain as the physiological basis for what makes human beings special or unique. If cells from humans and other animals can be mixed early in development and still form a functioning brain, then does it—will it—should it—matter what percentage of a brain is made up of human cells? Perhaps, instead, research will indicate that the timing of a genetic or cellular manipulation during organismal development is more important than the amount of material inserted? Whatever the case may be, brain experiments mixing cells from different species will certainly add to the challenges scientific research is raising to our commonly held concepts of what it means to be human, and what makes humans special—if anything.

In light of these challenges and the troubling ramifications they may have for our moral frameworks and ethical reasoning, due to their unsettling effects on our beliefs and concepts about human nature and human value, one could easily ask why it is that such research is being done at all. In order to answer this question well, at least a cursory understanding of stem cell research is required.

First of all, what are stem cells? The concept of stem cells is used to help explain how it is that a multicellular organism, such as a human being, can begin as a single cell and yet develop into a complex creature made of trillions of cells, that come in thousands of different types, which form hundreds of different tissues and organs, that provide the physiological basis for all our abilities and characteristics. In addition, many of the cells we require to function die during the course of a lifetime and need to be replaced. Stem cells are the source of these replacement cells. Hence, stem cells are considered to be special cells that can multiply to create and replace the many cells of our bodies, and at the same time replace themselves so that we continue to have some stem cells throughout our lives.[7]

From this understanding of stem cells, one can easily project several important goals for research using these amazing cells. Often these goals are grouped into three categories: basic research in human development, safer and more specific drug development, and therapies to repair or replace damaged tissues and organs.[8] The basic research is obviously significant because scientists want to understand better how human beings develop from a single cellular structure to the complex structure of an adult body. In addition, since stem cells function to replace the cells we lose in daily life, basic stem cell research may help answer questions about disease, injury, and aging.

The goal of safer and more specific drug development is one that might be less obvious to the public at large. The idea here is to use stem cells from different individuals to grow cells, tissues, or perhaps even organs. Then instead of, or in addition to, testing drugs on animals or generic human cell lines which may not represent accurately or precisely the reactions of a target human tissue, the cells or tissues grown from different individuals can be tested for the efficacy and toxicity of various drugs. From such research, companies might get a much better idea of which individuals would benefit more from which drugs, and which individuals should avoid which drugs, even before clinical trials with human subjects are begun.

Eventually, the goal is to develop products and therapies that would allow physicians to more directly, efficiently, and effectively replace and repair cells, tissues, and organs of an individual that may have been damaged or destroyed. This medical approach is now being promoted as "regenerative medicine."[9] In the public debate surrounding the stem cell issue, it is most often this goal of using stem cell research to regenerate tissues and organs that receives the greatest attention. Some additional distinctions concerning different types of stem cell research will help to clarify why this is the case.

The distinction most often used in the current stem cell debate is between "embryonic" and "adult" stem cells. "Adult" stem cells are something of a misnomer, since they are found in various tissues from the time of fetal development until death. "Embryonic" stem cells are those derived from the inner cell mass of a "blastocyst." "Blastocyst" is the term for a certain stage of human organismal development that is within the broader eight-week period of embryonic development. The blastocyst is a hollow sphere of cells with a cluster of cells inside. The embryonic stem cells are derived from this inner cluster, and are obtained by breaking open and thus destroying the blastocyst. Since the procurement of embryonic stem cells results in the destruction of embryos, this process is highly contentious in our society where many hold the position that human lives deserve protection from destruction for research purposes even, or especially, at this early stage of development.

If obtaining human embryonic stem cells is so controversial, then why would anyone want to do it? The answer to this question requires our returning to the above description of stem cells and human development. Since it is known that the human body begins development with the fertilization of an egg with a sperm, it can be concluded that all the different cells of an individual had their beginning in a single fertilized egg. Similarly, scientific evidence indicates that all the different cells of our adult bodies arise from some of the cells in the inner cell mass of the blastocyst. Using this information, researchers conclude that these embryonic stem cells must be able to make any cells or tissues one might need for research or therapy. Therefore, some researchers wish to use these embryonic stem cells to recapitulate what goes on during normal and/or abnormal human development.

Basically, then, the public debate concerning human embryonic stem cells revolves around weighing the good of doing this scientific research, with the primary goal of medical benefit, against the harms involved in doing research on human embryos. Having listed the benefits of this research above, I now turn briefly to the harms involved.

The most obvious, and probably the most broadly contentious, harm cited in the public debate is the destruction of the human embryo. This issue becomes exacerbated when proposals are made for intentionally creating human embryos, either by in vitro fertilization or by nuclear transfer techniques (cloning), in order to destroy them for their embryonic stem cells. At issue here is the value—moral, legal, social, etc.—societies are to give to or acknowledge in human embryos. The different arguments made concerning the value of human embryos range from claiming that they should be treated basically the same as any piece of human tissue to claiming that they should be treated basically the same as any human person. Since much has already been written across this broad spectrum, I wish to address only one aspect of the debate that highlights the uncertainty involved in this issue—the ambiguities encountered in this debate concerning the term "embryo."

In order to support the claim that human embryos should not have protections similar to those held by human subjects in general, it is often argued that the relatively high rate of embryo loss in early pregnancy (with some estimates at 50 percent[10]) indicates that embryos should receive a lesser moral and legal status than human subjects in general.[11] Otherwise, it is asked, why do not societies and cultures encourage the ritual mourning of the loss of these embryos, and why do they not advocate for greater medical interventions to save these human lives? Prescinding from an analysis of differing traditions concerning the appropriate response to death early in human development, one can, instead, evaluate the importance of this argument by focusing on the ambiguity, or even equivocation, inherent in this argument with respect to the use of the term "embryo."

When arguing about the ethical status of a human embryo, the underlying reality about which one is arguing can be described as that stage of human development we all transited on our way to our current stage of human development, whatever that may be. In other words, we are discussing human embryos in the context of what we ourselves once were. This context is not the same as the scientific one that undergirds the statistics about human embryo losses in early pregnancy. Such statistics might readily include abnormal growths, such as complete hydatidiform moles.[12]

Though hydatidiform moles may have characteristics similar to the embryos described above, these growths are not developing along the trajectory of a human organism. Rather, these growths are disorganized in their development to the extent that they may require surgical removal in order to prevent them from becoming life-threatening cancers. The question then arises: In light of the possibility of nonembryonic pregnancies, how many of these pregnancy losses are actually human embryos of the type of which we envision in our ethical debates? Once again, it appears that we are confronted with significant uncertainty. Since our scientific conceptualizations of an embryo may not match the embryo conceptualizations employed in our ethical analyses, the relevance of the argument regarding the percentages of embryos lost in pregnancy may be only minimal, at best, with respect to the human embryo research debate.

This problem of uncertainty in arguing about the ethical status of embryos fits within the larger context of uncertainty about human nature described earlier in this essay. It is not surprising that there is difficulty in defining the beginnings of human life, if it is indeed becoming more difficult to define human life itself due to our rapidly increasing scientific information. From this larger context, these uncertainties in the definition and understanding of embryos and human life may help to explain the impasse currently experienced in the human embryo research debates. If different, and even contrary, understandings of the beginnings of human life are being used in this public debate, then, without extensive clarifications, resolution of this contentious issue may be improbable, if not impossible. And if we cannot reach resolution on the status of the human embryo, how will we as a society address the coming dilemmas surrounding our concepts of human life or human nature?

I have argued elsewhere that the answers to these profound questions will require a revitalization of the philosophical anthropologies that undergird our ethical systems as well as our concepts of health, disease, and human nature.[13] This revitalization will likely entail broad interdisciplinary and intercultural dialogue, and, hence, some length of time. Still, as our society currently wrestles with these more fundamental questions, one needs to inquire what our society is doing now to address the debate concerning human embryonic research in spite of the contention and uncertainty surrounding this issue?

In one sense, this dilemma is not new for us, as we have already, as a society, decided that in light of past abuses such as the research performed on unknowing African-Americans or on the mentally disabled, it is sometimes best to limit what science and technology can do in order to better serve what is good for society.[14] In light of the harms caused to people in the name of scientific or medical progress, our society, and others around the world, have created guidelines and agencies to protect human research subjects from undue risks and harms.[15] This protection of human research subjects is an ongoing process with new revelations and investigations regularly being reported by government commissions and by the media.[16] The controversies surrounding human embryo research not only involve the debate over the status of human embryos, but also include other human subject issues such as the procurement of human eggs in large numbers, as might be required by nuclear transfer research and technology.[17] From within this current context of protections from undue research risk and harm, how is our current system of public ethical reflection responding to the human embryo research predicament?

One response to this contentious social issue has been for various organizations to gather panels of experts together to investigate, analyze, and evaluate the issue with the goal of generating recommendations for actions to be undertaken by governmental and/or other agencies. In general, the arguments and recommendations formulated by these expert panels have been reflective of, or employed by, many who are engaged in the broader public debate, especially with regard to legislation that has been or is being addressed on both the state and national levels. The arguments that have been made in support of human embryo research often fall into two primary categories, referred to here as arguments from "need" and "number."[18] A brief analysis of these arguments will reveal the uncertainties inherent in them, and, consequently, their insufficiency to serve as justifications for pursuing this socially contentious research.

Addressing the first argument, that of the need for human embryo research, it is important to recall that, as was observed in the beginning of this essay, the diseases suggested as likely targets for human embryonic stem cell research are also being targeted by researchers using other approaches, such as genetic therapies, drug development, and adult stem cells. It may well be the case that, for many patients, the treatments for their illnesses may come more quickly from research avenues other than human embryonic stem cell research and that these alternative treatments may even be better than any treatment derived from human embryonic stem cell research.

In response to this uncertainty as to what line of research might yet prove most successful in meeting the medical needs of people afflicted with severe or fatal diseases, proponents of human embryo research have argued that all scientifically sound lines of research should be pursued simultaneously so that we have the best chance of discovering what will work as soon as is possible. From a scientific perspective, this approach makes the most sense. In science, when there is uncertainty, one does all the research indicated to gain the desired knowledge and understanding. However, as was observed above, what is best for science is not always best for a society and its members. Some lines of research may be restricted or banned regardless of their scientific appeal in order to protect the well-being of a society. Research that is as controversial and contentious as human embryo research must have ethically compelling reasons to justify its pursuit if it is to carry any weight in countering the harm it creates.

At this point in the debate, human embryo research proponents often turn to the second argument cited above and emphasize the incredible number of people who could potentially benefit from such research. These proponents can point to the uncertainty inherent in all this biological research and argue that no society should deny all these people who suffer from severe and fatal diseases the potential benefits of this research, even if the research is controversial and contentious within a given society. Associating this research with the substantial societal value of medical healing gives this argument significance.

There is, however, a fundamental flaw in this argument that undermines its power and claim. The flaw in this argument lies in its assumption of a direct correlation between scientific or medical advance and medical benefit for those who need it. The realities of health care systems, both in our own society and around the world, argue against this assumption. With respect to health care in the United States, we need to acknowledge that, even if treatments from human embryonic stem cell research are the first to be proven successful, many, if not most, people who need these treatments will not get them.

Evidence of the accuracy of this bleak assessment of our health care system is found in the December 2001 report of the President's Cancer Panel. Though great strides have been made in cancer research during the past three decades of our war on cancer, the panel concludes: "In short, our health care system is broken, and it is failing people with cancer and those at risk for cancer—all of us."[19] Worldwide, the situation is much more bleak, considering that millions of children die each year from a lack of clean water, not to mention inadequate access to minimal health care technology.[20] Therefore, just because many people in the world might tragically share a devastating disease, such as diabetes or Parkinson's, one cannot conclude that this tragedy will be resolved by breakthroughs in research. The greater tragedy is that only a relative few get to enjoy the benefits of many of our medical research advances. The argument from number does not fit our social reality.

At this juncture it is critical that the arguments from uncertainty presented above be applied precisely. These arguments have been made to call attention to the flaws in the reasoning often presented in support of human embryonic research. These arguments do not argue against the pursuit of medical advances per se. These arguments do, however, place scientific and medical research in the larger context of the good of societies in general. The National Bioethics Advisory Commission acknowledged the importance of this context and the consequent requirement for greater justification than normal in pursuing scientific research that is socially contentious.[21] Therefore, if the justification for proceeding with the destruction of human embryos for research rests even in part on these claims of need and number, then this justification is flawed and requires rethinking.

The evidence and analysis put forward in this essay attest to the pervasiveness of uncertainty in all aspects of the human embryo research issue. This uncertainty, it has been argued, even undermines the proposals for pursuing this research put forth by some of the expert panels commissioned to address this issue.

How then should society proceed? The arguments of this essay suggest two responses that could be implemented immediately within the current circumstances of our society. First, in recognition of the need for research into stem cell biology in order to understand better its promises and perils for future societal decisions, governmental support should be increased for stem cell research using animal models and nonembryonic human stem cells. This response would achieve scientific progress without raising a host of social and ethical concerns. Second, in recognition of the vast numbers of people, within our own nation and around the world, who suffer from severe and lethal diseases or injuries, the findings and recommendations for improving health care proposed by expert groups, such as the President's Cancer Panel and the World Health Organization, should receive the same level of attention and action as has been expended on human embryo research.

NOTES

1. The National Center for Biotechnology Information has websites that provide an overview of many of these technologies and has internet links to other resources explaining these technologies. See both www.ncbi.nlm.nih.gov/About/primer/index.html and www.ncbi.nlm.nih.gov/About/outreach/index.html.

2. See Nikunj Somia and Inder M. Verma, "Gene Therapy: Trials and Tribulations," *Nature Reviews Genetics* 1, no. 2 (November 2000): 91–99; and D. Teichler Zallen, "US Gene Therapy in Crisis," *Trends Genetics* 16, no. 6 (June 2000): 272–75.

3. F.S. Rosen, "Successful Gene Therapy for Severe Combined Immunodeficiency," *New England Journal of Medicine* 346, no. 16 (18 April 2002): 1241–43; and S. Hacein-Bey-Abina, F. Le Deist, F. Carlier, C. Bouneaud, C. Hue, J.P. De Villartay, A.J. Thrasher, N. Wulffraat, R. Sorensen, S. Dupuis-Girod, A. Fischer, E.G. Davies, W. Kuis, L. Leiva, and M. Cavazzana-Calvo, "Sustained Correction of X-Linked Severe Combined Immunodeficiency by Ex Vivo Gene Therapy," *New England Journal of Medicine* 346, no. 16 (18 April 2002): 1185–93.

4. Gene replacement could involve techniques known as homologous recombination, while gene repair could be done using techniques that correct a mutation by replacing the single letter (nucleotide) that is misspelled in the DNA. For example, see P.D. Richardson, L.B. Augustin, B.T. Kren, and C.J. Steer, "Gene Repair and Transposon-Mediated Gene Therapy," *Stem Cells* 20, no. 2 (March 2002): 105–18.

5. H.F. Willard, "Artificial Chromosomes Coming to Life," *Science* 290 (2000): 1308–9.

6. V. Ourednik, J. Ourednik, J.D. Flax, W.M. Zawada, C. Hut, C. Yang, K.I. Park, S.U. Kim, R.L. Sidman, Curt R. Freed, and Evan Y. Snyder, "Segregation of Human Neural Stem Cells in the Developing Primate Forebrain," *Science* 293 (7 September 2001): 1820–24.

7. For background on stem cell biology and the perspective of the National Research Council and Institute of Medicine committee on stem cell research, see their report, *Stem Cells and the Future of Regenerative Medicine* (2002), which can be found online at www.nap.edu/books/0309076307/html.

8. See the NIH stem cell primer webpage, www4.od.nih.gov/stemcell/figure5.jpg.

9. National Research Council and Institute of Medicine, *Stem Cells and the Future of Regenerative Medicine.*

10. For more on miscarriage and its causes, see the March of Dimes website at www.modimes.org/HealthLibrary/334_592.htm.

11. For two examples of this type of argument, see Jeffrey Spike, "Bush and Stem Cell Research: An Ethically Confused Policy," *American Journal*

of Bioethics 2, no. 1 (2002): 45; and Robert Baker, "Stem Cell Rhetoric and the Pragmatics of Naming," *American Journal of Bioethics* 2, no. 1 (2002): 53.

12. For more on molar pregnancies see www.modimes.org/Health Library/334_591.htm.

13. "Philosophical Anthropologies and the HGP," in *Controlling Our Destinies: The Human Genome Project from Historical, Philosophical, Social, and Ethical Perspectives,* ed. Phillip R. Sloan (Notre Dame, Ind.: University of Notre Dame Press, 2000).

14. Adil E. Shamoo and Joan L. O'Sullivan, "The Ethics of Research on the Mentally Disabled," in *Health Care Ethics: Critical Issues for the 21st Century,* ed. John F. Monagle and David C. Thomasma (Gaithersburg, Md.: Aspen Publishers, 1998).

15. Guidelines for protecting human research subjects are found in such reports as the World Medical Association's "Declaration of Helsinki," and the U.S. "Protection of Human Subjects," *Code of Federal Regulations* 45, CFR 46, rev. 8 March 1983.

16. Examples of current events include the Virginia governor's apology for forced sterilizations (www.governor.state.va.us/Press_Policy/Releases/May2002/May0202.htm) and recent Navy bioweapon exposures reports (www.cbsnews.com/stories/2002/05/24/national/main510079.shtml).

17. Andrew Pollack, "Use of Cloning to Tailor Treatment Has Big Hurdles, Including Cost," *New York Times,* 18 December 2001, F2.

18. Though many examples of these basic arguments can be found in the transcripts of congressional and state legislature hearings, one can find the arguments from need and number clearly stated in the following documents from three of the highest profile expert panels assembled to date: American Association for the Advancement of Science and the Institute for Civil Society report, *Stem Cell Research and Applications: Monitoring the Frontiers of Biomedical Research* (www.aaas.org/spp/dspp/sfrl/projects/stem/report.pdf); National Bioethics Advisory Commission report, *Ethical Issues in Human Stem Cell Research,* 3 vols. (September 1999) (www.bioethics.georgetown.edu/nbac/pubs.html); and the National Research Council and Institute of Medicine report, *Stem Cells and the Future of Regenerative Medicine.*

19. President's Cancer Panel report, "Voices of a Broken System: Real People, Real Problems" (December 2001), 2 (www.deainfo.nci.nih.gov/ADVISORY/pcp/video-summary.htm).

20. World Health Organization, "Children's Environmental Health" (www.who.int/peh/ceh/index.htm).

21. NBAC *Ethical Issues in Human Stem Cell Research,* 1:52–53.

CHAPTER 3

Stem Cell Research and Religious Freedom

JOHN LANGAN, S.J.

I would like to consider the morality of stem cell research not as a first-order issue in which the crucial question is whether such research conducted on embryos is morally right or wrong, but as a second order question about the way in which it is justifiable for us to treat such research in a pluralistic society—and more specifically about whether our commitment as a church to religious freedom has any implications for this question as it arises within the domain of public policy. I should begin by stating that I am not an expert on stem cell research or even on bioethics more generally conceived. The main focus of my concern over the last few years has been Catholic social thought; when questions of bioethics arise, my colleagues in the Kennedy Institute are ready to concur in my judgment that I am learner rather than teacher. I should take the further step of stating that I believe that there is a strong prima facie case against experimental research on embryos.

It will not be my task to subject that view to critical scrutiny or to defend it; rather, I will assume that that view stands as a conclusion from natural law principles and from a nondualistic interpretation of the process of human fetal development. Catholic doctrine in this area rests on rationally accessible propositions about the substantial continuity of the person from conception on to maturity and about the respect that should be shown for innocent human life in general. In my understanding of it, it is not a view that relies on religiously idiosyncratic premises or that requires the gift of faith for its affirmation. It may well be the case in our society that only Catholics and conservative evangelical Protestants deny the moral acceptability of using embryos for experimental purposes. It does not follow from this fact that the conclusion is intrinsically religious. But it remains true that the conclusion is embedded with a context of religious controversy and is often presented or denounced as a purely religious view.

We need to ask ourselves where this conclusion about the unacceptability of using embryos in this way fits within the landscape of natural law theorizing about morality. Here I think that our best guide is the classical presentation of the precepts of natural law found in Thomas Aquinas, *Summa Theologiae,* I–II, question 94, especially in articles 2 and 4.[1] Thomas here is considering the questions of whether the natural law contains one precept or many and whether natural law is the same for all, both with regard to our knowledge of it and with regard to the requirements for rightness of action that it imposes on us. If I may bring together what Aquinas says in these two places (as well as what he says about the commandments in question 100), it seems to me that Aquinas thinks that natural law precepts are found on three levels of specificity and knowability. Briefly, these are:

(a) highly generic precepts such as "Good is to be done, evil is to avoided," as well as the love command. These, he tells us, are known to all who understand the meaning of the key terms. A contemporary philosopher could add that they would be denied only by those who suffer from such a failure of linguistic and moral understanding that they lack the capacity to enter into the linguistic and moral community.

(b) specific precepts forbidding certain kinds of actions, e.g., the prohibition of killing in the Ten Commandments. These precepts correspond to the natural inclinations enumerated in question 94, article 2; that enumeration breaks natural inclinations into those we share with all beings (such as to continue existing), those we share with animals (inclination to sexual union of male and female), and those that are distinctively human (to live in society and to know the truth about God). The precepts corresponding to these inclinations are known to all, though Aquinas is quite vague on precisely how they are related to the natural inclinations and on how they are to be given social recognition and enforcement. They are somehow derived from or are at least consistent with the generic precepts; but the key point for our purposes is that they are known to all without complex or sophisticated argument. Simple deduction or intuition seem to be the main plausible proposals for how we know them. Denial of one or more of these principles would put one outside what Alan Donagan has discussed as "Hebrew-Christian morality" and what others have treated under the heading of "common morality."[2] Aquinas attributes the failure to acknowledge these precepts to "a reason depraved by passion or by evil custom or from a bad cause of nature, as theft was once thought by the Germans not to be wrong, although it is expressly against the law of nature" (*S.T.* I–II, q. 94, a. 4c). A failure to acknowledge these secondary precepts creates a social division between those who acknowledge them and those who do not; it has a community-shattering effect. This should, of course, be distinguished from the less radical disruptions produced by nonobservance of the precepts.

(c) further conclusions and specifications of these precepts, which would, in Aquinas's terminology, be known to the wise or the learned (*doctis,* literally those who have been taught). These further conclusions, such as the negative judgment on stem cell research, require additional premises about the development of human life, and do not follow merely from the generic and specific precepts. Here it should be remembered that Aquinas is talking about a logical ordering of precepts, not about a historical order of discovery or about the way in which a society comes to recognize an evil or a moral obligation.[3] Our knowledge of various moral values and norms has not developed in a purely deductive fashion, but it moves forward mainly in response to the recognition of moral and social problems. We are often more convinced that some course of action is wrong than we are able to provide clear and compelling reasons why it is wrong.[4]

If something like this classification is correct, then we have the problem of how to deal with disagreement, especially at the third level or with regard to what have often been discussed as the tertiary precepts of the natural law.[5] In particular, how should we deal with what appear to be irreconcilable differences with our fellow citizens who seem to be making conscientious judgments and who describe and analyze the situation in different terms than we do and reach significantly different judgments? The crucial question is whether we should attempt to use the coercive power of the law to require obedience to norms that are opposed by these conscientious judgments. Granting this point does not require us to attribute good faith to all proponents of stem cell research, much less to all proponents of abortion. We can still contend that "the culture of death" has had a negative influence on the moral perceptions and judgments of large numbers of our fellow citizens.[6] We can still be dismayed that so many people ignore the scientific and medical evidence for the continuity of the developing human being in utero. But we need to be seeking for greater clarity about the basis on which we propose to require that others accept our conclusions as a guide for public policy in a matter that cannot be settled by the invocation and repetition of commonly accepted principles of the types found on the first two levels.

The tendency within Catholic and conservative Christian discourse has been to treat advocates of stem cell research as people who fail to recognize the sanctity of life, that is, in effect, as people who reject precepts of the second type. My suggestion is that we should be treating them as persons many of whom have reached different conclusions with regard to precepts of the third type. The reason for this difference is often not so much a conflict of values as it is reliance on different models or accounts of the problem. Rejection of precepts of the second type presents a threat to the stability of the community, whereas disagreements about the extent and application of precepts of the third type are compatible with the continued flourishing and civic peace of the community, provided that they are conducted in a way that manifests mutual respect.

Political and psychological realism impels us to recognize that ordinary citizens will not, in practice, treat embryos and newborns and adults in the same fashion and will not react to the termination of their lives in the same way. In particular, they will be ready to accept a wider range of justifying reasons for terminating life in the earlier stages of the developmental process. Many of these reasons will be morally weighty in themselves, such as potential benefits to the victims of various illnesses; and so we can acknowledge that there is some morally worthy basis for the conclusions even of those with whom we disagree. We also have to acknowledge that many of those who disagree with us do so for reasons that they regard as imposing requirements of conscience and as setting a direction for their own exercise of religious freedom.

The vigorous affirmation of the continuity of human life in all its stages, which has been characteristic of Catholic teaching over the last century, alerts us to the inconsistency commonly found in those who try to compromise the differences of principle that arise both in the abortion debate and in the stem cell debate. But it also runs the risk of encouraging those who dream of resolving these matters on an all-or-nothing basis. For consistency taken in isolation makes an ambiguous contribution to outcomes in this area; it may push people to accept things such as late-term abortions and infanticide, which they had previously rejected, and to reject things they had previously accepted, such as the sanctity of life in the later stages of fetal development. A gain in consistency may well be accompanied by a decline in moral acceptability. It may be that, in a complex world marked by vigorous disagreement about tertiary precepts and by a distressing lack of clarity about how to proceed from primary and secondary precepts, the values and the lives we are concerned to protect are safer in a resolution of the problem that may lack consistency but which is likely to draw conscientious support from a broad range of people in the civic and political community.

Given that the stem cell debate is unlikely to resolve itself into a broadly shared consensus, and given that its progenitor, the abortion debate, will continue to be highly divisive, what conclusions follow for public policy and public debate from the considerations I have been offering? More specifically, what conclusions can we discern for those Catholics who believe that abortion is wrong and that embryonic stem cell research, despite the good intentions of its advocates, is morally tainted and who believe at the same time that the shared respect for religious freedom and for freedom of conscience is one of the core values of our society? Please note that I speak of "discerning" these proposed conclusions. In using this language, I mean to present these proposals as guidance for a conflictual situation in which it does not seem possible to achieve all the relevant and necessary goods in a harmonious way. I also mean to make it clear that I do not regard our resolution of this and similar conflicts as being simply the application of a universal principle to a particular case or set of cases, but as arising from attentive reflection on the various claims and moral aspirations voiced in the debate.

The conclusions I propose are briefly three:

1. Catholic institutions ought not to participate in embryonic stem cell research.
2. Catholics should be prepared to tolerate the carrying forward of such research by those committed to it, even while raising critical questions about it.
3. Government agencies should refrain from funding such research out of respect for the consciences and the religious freedom of those who find it deeply troubling.

The first point is open to criticism by those who believe that a morally acceptable way of engaging in embryonic stem cell research can be found. This amounts to a resolution of what I referred to earlier as the "first-order debate" about the morality of stem cell research. My comments here have not dealt with the difficult issues of that debate, and I am both heartened and discouraged to find persons whom I greatly respect on both sides of the debate. If a way should be found to establish the acceptability of embryonic stem cell research within the Catholic moral tradition, and if this position should find acceptance within the Catholic moral community, then the difficulty is resolved; and there would be no need for Catholic institutions to refrain from such research. The first point is intended for those scenarios in which such a resolution of the problem does not achieve intellectual or ecclesial acceptance or in which the matter is still regarded as highly contentious. We would then be dealing with a prohibition of research that rests on a particular way of interpreting and specifying the precept against the taking of innocent human life. Given the seriousness and depth of the church's commitment to the protection of innocent human life, it seems to me that Catholic institutions should accept such a prohibition, which arises from a concern for the integrity of Christian witness in this vital area. This does involve a painful subordination of promising research agendas, a subordination difficult both for the scientists who are interested in pursuing the research and for the university leaders, both administrators and faculty, who are anxious to affirm and protect the value of academic freedom in Catholic institutions. Here it seems to me that the important step is to perceive and to present this prohibition as one of many areas in which our society and its communities recognize the importance and value of ethical restraints on research agendas. The vast body of regulations and professional literature on the protection of human subjects should remind us that we are not dealing with yet another episode in the warfare of theology and science, but rather with another decision of a type that should concern all thoughtful members of our society and that should remind us that we will not be helped at all by ways of thought which rely on simple polarities and stark antagonisms.

The second point counsels those who regard embryonic stem cell research as morally tainted to take seriously the views of those who differ with us and to treat the persons who hold these views with respect and with charity. There is and can be no straight line from mistaken moral judgment to radical defects of character. Disagreements of the sort we encounter in the stem cell research debate are, given Aquinas's account of how we come to know the precepts of natural law, what we should expect to find. In particular, we should avoid imputing moral and intellectual dishonesty to those who disagree with us and should avoid ascribing to them a denial of the worth of human life. We should engage with their ideas and we should assess their scientific research critically and professionally.

The third point is more in the nature of a general moral or an analogy drawn from the history leading to separation of church and state. It is a point I make to the proponents of embryonic stem cell research rather than to the opponents. It seems to me that few things are more likely to produce strident conflict than government funding not merely of religious groups but also of practices and programs that strike significant parts of the political community as morally objectionable. Among those secular liberals most concerned with the First Amendment, there is a common tendency to assume that the threats to religious freedom and to freedom of conscience arise from religious institutions and their constituencies. There is also a tendency to overlook the ways in which government regulations and secular expectations can be experienced by religious people as an interference with or burden on their consciences. In a free society, proponents of embryonic stem cell research have abundant opportunities to educate people on the merits of what they propose to do and to make requests for government funding. They also have the freedom to carry on their research without government funding. If their position is correct, and if the research they wish to pursue promises significant benefits, and if embryonic stem cell research is the uniquely necessary way to achieve these benefits, then it seems that they have a strong case for claiming the further advantages provided by government funding. But if their persuasive efforts fail to build a strong national consensus in favor of federal support of embryonic stem cell research, and if large groups in the general population make it clear that they continue to have conscientious objections to such research, then the proponents of embryonic stem cell research should be ready to acknowledge that an important good can still be realized: namely, the good of showing respect for the consciences of those who oppose them. This is a good that is essential for a pluralistic democracy, and it is often lost sight of in the heat and dust of controversy.

NOTES

1. See Thomas Aquinas, *Summa Theologica* vol.1, trans. Fathers of the English Dominican Province (New York: Benziger Brothers, 1948).

2. See Alan Donagan, *The Theory of Morality* (Chicago, Ill.: University of Chicago Press, 1977).

3. For a more sustained argument that Aquinas offers something beyond a purely deductive account of morality, see John Langan, "Beatitude and Moral Law in Thomas Aquinas," *Journal of Religious Ethics* 5 (1977):183–95.

4. For a critique of deductivist accounts of moral reasoning in contemporary Catholic theology, see John Langan, "Moral Rationalism," *Theological Studies* 50 (1989).

5. An accessible and straightforward presentation of the division into primary, secondary, and tertiary precepts or principles, a division that runs through many neoscholastic manuals of moral philosophy and theology, is given in Thomas Higgins, S.J., *Man as Man* (Milwaukee: Bruce, 1958), 120–22.

6. This expression figures prominently in the important encyclical of John Paul II, *Evangelium Vitae* (Washington, D.C.: U.S. Catholic Conference, 1995), pars. 12 and 28.

CHAPTER
4

Umbilical Cord Blood, Stem Cells, and Bone Marrow Transplantation

RONALD M. KLINE

Pluripotent stem cells (PSCs) have the potential to revolutionize the future of medicine because now, for the first time, we can envision the ability to repair and replace those complex organs and tissues whose failure currently leads to the disability and premature demise of millions of people. Unfortunately, the acquisition of PSCs from early stage human embryos forces our multicultural society to attempt to reach a consensus about what constitutes human life, and at what point during embryonic development that life begins. Umbilical cord blood (UCB) is a potential alternative source of PSCs, bridging that ethical divide by providing a PSC source from a biological product that is today, with rare exceptions, simply discarded as biomedical waste.

UCB emerged in the early 1990s as a revolutionary source of hematopoietic stem cells (HSCs) that promised to dramatically improve the perennial problem of blood and marrow transplantation (BMT): the search for tissue-matched donors for patients needing an allogeneic BMT (that is, BMT from a nonself donor). Recent discoveries in the field of stem cell research may add to the already remarkable potential of UCB by promising to make this abundant by-product of birth a source of PSCs able to treat a wide array of diseases, while continuing in its more established role as a source of the more differentiated HSCs that can reconstitute the marrow compartment of a human being.

Although UCB could provide a near limitless source of PSCs, with four million births a year in the United States alone, the advantages, disadvantages, and limitations of this PSC source compared to those derived from adult and embryonic sources remains to be fully defined in this young and rapidly progressing scientific field. In addition, the ethical concerns raised about UCB in the BMT setting also apply to its use as a source of PSCs as well, and thus are relevant to the ongoing debate about clinical applications of stem cell research.

UMBILICAL CORD BLOOD AS A SOURCE OF HEMATOPOIETIC STEM CELLS

Bone marrow transplantation was first successfully reported in 1968 in two children with immune deficiencies.[1] The successful reports followed more than a decade of failures in patients with a wide variety of diseases. In the thirty-five years since these first successful transplants, the field of bone marrow transplantation has advanced considerably, and BMT has been used to treat a variety of malignancies, hemoglobinopathies (such as thalassemia and sickle cell disease), immune deficiencies, and congenital metabolic defects. It has also emerged as a promising new therapy for a broad range of autoimmune diseases such as lupus, scleroderma, and multiple sclerosis. Nevertheless, BMT has been hindered since its first successful clinical applications in the late 1960s with the need to find "matched donors." For a BMT to have the greatest chance of success, the major transplantation antigens of both donor and host must be fully matched. These proteins exist in all vertebrate species and are referred to in general terms as the major histocompatibility complex (MHC). In humans they are termed human leukocyte antigens or HLAs and are genetically encoded on chromosome 6. In clinical BMT, three HLA gene products (HLA-A, HLA-B, and HLA-DRB1) have been identified as having clinical relevance. Since both a maternal and paternal allele exist for these proteins, six HLA gene products are typed in preparation for BMT and a full match is termed a "six out of six" match. Mendel's Law of Independent Assortment dictates that the possibility that a sibling will inherit the same maternal and paternal chromosome 6 as his sibling in need of a transplant is 25 percent (50 percent chance of inheriting the same maternal chromosome 6 and 50 percent chance of inheriting the same paternal chromosome 6). Taking into account the size of the average American family, a patient needing a BMT has an approximately 30 percent chance of having an HLA match in a sibling.[2] Thus, relying on HLA identical siblings as the sole donor source leaves seven out of ten potential transplant patients without the possibility of curative therapy.

As a partial remedy to this problem, HLA-matched unrelated donor transplants were first attempted in 1973,[3] and many successful transplants were subsequently reported.[4] As a direct result of these efforts, the National Marrow Donor Program (NMDP) was founded in 1986 to facilitate unrelated donor transplants in the United States. Similar registries have been established worldwide. The NMDP currently has over four million U.S. donors in its data banks.[5] When combined with the thirty-nine international registries, over six million unrelated donors are available to patients in need of a donor.[6] Nevertheless, the NMDP is successful in identifying a donor only 75 percent of the time. The percentage is even less in minority patients, where underrepresentation of donors decreases the chances of finding a matched donor.[7]

Other alternatives to matched related and unrelated transplants include T cell–depleted bone marrow transplants and autologous transplants (using one's own marrow as a source of HSCs). Both of these alternatives, as well as the use of unrelated donors, have significant additional risks associated with them—risks not present in matched sibling transplants—and, therefore, they have not been a universal solution to the problem of donor scarcity.

Over the last decade, a new source of HSCs has become available with the potential to significantly alleviate the shortage of donors that has plagued bone marrow transplantation since its inception. Beginning in the early 1980s, it was demonstrated that UCB contained high levels of hematopoietic progenitor cells,[8] with a report in 1989 from Broxmeyer et al., demonstrating that the numbers of colony-forming units (an in vitro indicator of engraftment potential) contained in UCB collections was similar to that obtained from marrow collections where sustained hematopoietic engraftment had been achieved.[9] This data followed two case reports from the late 1960s and early 1970s suggesting that cord blood infusions caused transient changes in RBC (red blood cell) phenotype, not related to the infusion itself, when administered in the clinical setting of conventional dose chemotherapy.[10]

The first umbilical cord blood transplant with sustained engraftment was performed in 1988 on a child with Fanconi's anemia, the transplant coming from his HLA identical sibling.[11] The child continues to do well to this day. This initial demonstration of the effectiveness of UCB in providing hematopoietic engraftment rapidly generated tremendous clinical activity centered on determining the proper uses of this virtually limitless supply of HSCs.

CLINICAL EXPERIENCE

Since the first successful cord blood transplant in 1988, the field of umbilical cord blood transplantation has expanded dramatically with several thousand transplants worldwide since 1988. It is estimated that more than 75 percent of these transplants have used unrelated donors.[12] Initial concerns centered on the small number of mononuclear cells infused, which are generally one-tenth of the number of cells required for engraftment in more traditional forms of transplantation. These concerns have been lessened by clinical experience with umbilical cord blood transplantation, which has demonstrated successful engraftment in both pediatric and adult recipients. Nevertheless, data does show more rapid engraftment in smaller recipients of umbilical cord blood transplants.[13]

UMBILICAL CORD BLOOD AS A SOURCE OF PLURIPOTENT STEM CELLS

At this time, UCB has not been shown in either animal or human models to be a source of PSCs. It is clearly a source of HSC, and thus the potential of UCB must largely be extrapolated from data using marrow cells. These cells have been shown in preclinical models to differentiate into skeletal and cardiac muscle, hepatocytes, vascular endothelium, and neural tissue.[14] Clinically, allogeneic BMT has been shown to improve osteogenesis imperfecta, a defect in the mesenchymal cells that produce Type I collagen matrix.[15] It is not clear from either the preclinical or clinical models if this differentiation into nonhematopoietic tissues represents the dedifferentiation of HSCs into PSCs, or if PSCs exist alongside HSCs in the marrow compartment. If the former proves correct, then UCB will also likely be a source of PSCs. If the latter is the case, then these PSCs must also exist alongside HSCs in UCB for the potential of UCB to be realized.

At this writing, the field of stem cell research is still too undeveloped for a detailed analysis of advantages and disadvantages of one source of stem cells versus another. Significant advances in this young field will undoubtedly clarify the picture over time and simplify what now appears to be a bewildering array of alternatives. It is also possible, as has been the case for BMT, that different stem cell choices will have different advantages and disadvantages, mandating their choice in different clinical scenarios.

HOW CORD BLOOD IS COLLECTED

Collection of UCB is a technically simple procedure that poses no foreseeable health risks to either mother or baby. The most widely used approach is to wait until the placenta is delivered and then to place the placenta in a sterile supporting structure with the umbilical cord hanging through the support. The umbilical cord is then cleansed with Betadine and alcohol, and the umbilical vein is accessed using a standard blood collection needle connected to a standard blood collection bag with anticoagulant and nutrient solution. UCB is then collected by gravity drainage, yielding approximately three ounces of blood.[16] It is then cryopreserved using standard HSC techniques. In a variation on this procedure, UCB is red-cell depleted prior to cryopreservation, thus both reducing the storage volume to approximately one ounce (important when large-scale cord blood collection is envisioned) and eliminating issues of blood type and Rh factor compatibility at the time of marrow infusion.[17]

An alternative method involves collecting the UCB after the delivery of the child while the placenta is still in utero (the third stage of labor). Such a technique has the theoretical advantages of beginning collection earlier, before coagulation within the placenta can begin, as well as using the contractions of the uterus to enhance blood collection. These advantages are theoretical at this time, since no large comparative studies have been published. Certainly, this latter technique is more intrusive and has the potential to interfere with the mother's care after delivery.

POTENTIAL ADVANTAGES OF UMBILICAL CORD BLOOD

It appears likely that HLA matching will have just as important a role in the transplantation of PSC-derived tissues as it does for BMT. This is because the PSC-derived tissues will express human leucocyte antigens on their surface, causing them to be rejected by the host immune system if not sufficiently matched. Therefore, many of the clinical issues relevant to the use of UCB for BMT may apply to the clinical applications of PSC as well.

Size of the Potential Donor Pool. It has taken the NMDP over fifteen years to accumulate a donor pool of four million individuals. This number represents the total births in the United States in just a single year. Thus, the rapid accumulation of enough UCB samples to provide for anyone needing tissue-matched allogeneic PSC is truly an achievable goal. As an example, the New York Blood Center can provide full– or one-antigen mismatched donors for approximately half of its requests using stored blood from a pool of only 16,000 umbilical cords.[18]

Speed. Identifying a suitable non–cord blood, unrelated donor is a time consuming process, taking an average of four months from search initiation to marrow delivery.[19] During this period, potential donors go to donor centers to have blood drawn for confirmable high resolution HLA typing and viral testing. After a donor is selected from this pool, that individual must return, pass a physical examination, and then schedule a bone marrow harvest. In contrast, cord blood has undergone viral testing upon storage, and cryopreserved DNA samples are available on-site for confirmable high-resolution HLA testing. Thus an umbilical cord blood transplant can be facilitated in as little as a few days.[20] For those patients for whom the acquisition of PSC-derived tissues is time critical (a patient with heart failure after a major heart attack, for instance), the rapidity with which UCB can be accessed can be lifesaving.

Racial Diversity. HLA phenotypes tend to segregate in racial groups, making it more likely that a suitable donor will come from the same racial group as the recipient. NMDP statistics show that a Caucasian will find a match 81 percent of the time, while the corresponding probabilities for African-Americans are 47 percent, Hispanics 64 percent, Pacific Islanders/Asians 55 percent, and American Indians/Alaskan Islanders 75 percent.[21]

In some cases, this represents increased HLA diversity in some ethnic groups as compared to Caucasians. For example, because African-Americans originate from the geographic area where Homo sapiens evolved, rather than a population subset that migrated to other continents (Europe, Asia, etc.), they have more HLA diversity in their population (and thus a more difficult time locating a suitable donor) than other ethnic groups.[22] In addition, the genetic mixing that has occurred between the African-American and Caucasian populations during three centuries in North America adds still more diversity to African-American HLA phenotypes and makes it even more difficult to find suitable donors for these patients.[23]

UCB harvesting can overcome these limitations both for BMT and for PSC applications by focusing collection efforts in hospitals where the children of underrepresented ethnicities are born.

POTENTIAL DISADVANTAGES

Transmission of Genetic Diseases. Cord blood harvested from the fetus represents an untested source of hematopoietic stem cells: "untested" in that the fetus has not yet demonstrated his health and viability in the external environment over several years, as is the case for other donors. Therefore, it is possible that congenital diseases, clinically unapparent in the fetus at birth, may transmit disease to recipients via the tissues derived from their PSC. These diseases could range from benign to life threatening, depending on the disease and the tissue type involved.

Many of these congenital abnormalities could be detected by the active clinical follow-up of cord-blood donors six to twelve months after birth. This would, however, require the creation of a long-term identification link between a donor and his cord blood unit, as well as continued contact with the donor center. The prospect of such a linkage has created serious privacy concerns among medical ethicists and is currently the subject of active debate.[24] In its place, collection centers have potential donors complete detailed questionnaires prior to UCB collection, with particular emphasis on individual and family histories of disease, as well as a detailed sexual history. If responses on the questionnaire generate medical concern, then the unit is not collected.[25]

Clearly, the potential for transmission of genetic diseases exists for embryonic stem cells, as well as UCB, although it would not be the case for PSC derived from adult donors.

Long-term Storage. Currently, there is limited data on the viability of UCB in long-term liquid nitrogen storage. The longest that a cord-blood sample has been cryopreserved and then successfully used for BMT is eight years.[26] No one yet knows the limits of cord-blood viability in liquid nitrogen storage. As an approximation, it is known that cryopreserved autologous bone marrow stored for greater than two years has allowed successful engraftment in 94 to 97 percent of patients.[27] In one case, the marrow had been stored for eleven years.[28] Whether these findings can be generalized to PSC applications is unknown, and yet viability is critical to the success of all cord-blood storage efforts, both for BMT and the clinical applications of PSCs. If, for example, UCB is not viable in liquid nitrogen storage for the years it takes a person to move from infancy to being an elderly adult afflicted with the diseases for which PSC technology is currently envisioned, then clearly storage of one's own UCB for future use would not be a meaningful option. Conversely, this would not affect the storage of UCB in donor banks where they were constantly used and replenished.

ETHICS

The availability of UCB collection efforts has raised many ethical issues apart from those of more traditional forms of BMT.[29] As in many other types of organ transplant, the ethical issues revolve around those of ownership, privacy, and allocation of limited resources. When applied to the clinical applications of PSCs, the issues remain largely similar.

Questions have been raised as to whether a UCB donor is entitled to reclaim his donated cord blood in the event that he or a relative needs it, and whether he is entitled to a share of the fees charged by collection banks for the UCB. This has raised UCB ownership issues, as well as the ever-present privacy concerns created by the permanent identification record that would be required. As UCB becomes a more valuable resource with the potential for stem cell generation, the above issues will only become more difficult.

In addition to unrelated UCB banks, several organizations have begun to offer, as a "for profit" service, the cryopreservation and storage of UCB. The UCB would therefore be available to that individual at a later time should he develop a condition warranting umbilical cord blood transplantation or a need for pluripotent stem cells. The largest of these companies, ViaCord, Inc., in Boston charges $1,550 for the initial cryopreservation, with an annual fee of $95 for continued storage.[30] This facet of UCB storage has raised ethical concerns about the potential availability of lifesaving technology based on economic means. To the extent that UCB proves to be a source of PSCs that have the potential to cure diseases later in life, this argument will only become stronger and more troubling.

Conversely, questions have also been raised about the ethics of marketing an expensive service to new parents when the probability of actually needing an autologous cord-blood transplant ranges upwards from one in ten thousand. Ownership issues have also been raised in the context of this service as well. Does the UCB belong to the child from whom the placental blood was taken or to the parents who presumably fund the cryopreservation and storage fees? This issue becomes relevant if the parents wish to use the UCB for a purpose other than autologous transplantation, such as for allogeneic transplantation into a sibling. In other words, do the parents of a minor child have the right to use the UCB in the best interest of a sibling (or cousin) rather than keeping the UCB cryopreserved indefinitely in case of need by the donor? Do parents have the right to sell the cord blood in a case of financial hardship? Finally, what are the rights and obligations of a storage facility if the storage fees go unpaid? Does it have the right to sell or otherwise dispose of the cord blood? Once frozen, does it have an ethical obligation to keep the cord blood in storage in perpetuity, regardless of whether or not storage fees are paid?

CONCLUSIONS

The discovery of PSCs has raised the hopes of millions of people afflicted with a wide range of diseases for which there is either no cure or a very limited one. On the other side of the argument are those who view the embryo as a human being with all the rights of an independent person. Complicating the argument still further has been the discovery of multiple sources of PSCs, with the advantages and disadvantages of each source the subject of much speculation but little data. Clearly, this knowledge will be crucial to the ongoing debate over the sources and clinical applications of PSC technology. If, for example, adult-derived stem cells prove to be as malleable as embryonic or hematopoietic stem cells, then every human will have his own fully matched reservoir of stem cells readily available, and much of the preceding discussion will be moot. If, however, embryo-derived PSCs prove to be advantageous because of their relative lack of senescence in comparison to adult-derived PSC, then cord blood–derived PSCs may prove to be a medically and ethically acceptable alternative that will bridge the gap between the two sides of this passionate argument.

The future may bring a world in which each individual has his UCB stored at birth in anticipation of a future time when it will be used to heal a failing organ. Conversely, we may see an expansion of the mission of currently existing government-funded cord-blood banks to include such uses as the provision of HSCs for BMT and the provision of PSCs for the treatment of innumerable, previously incurable diseases. Lastly, all of the above may be irrelevant as adult-derived stem cells prove to be the equivalent of other sources, allowing each of us to repair our bodies from the stem cells residing in our own marrow compartments or adipose tissue.

NOTES

1. See R. A. Gatti et al., "Immunological Reconstitution of Sex-Linked Lymphopenic Immunological Deficiency," *Lancet* 2 (1968): 1366–69; and R. J. Albertini et al., "Bone Marrow Transplantation in a Patient with the Wiskott-Aldrich Syndrome," *Lancet* 2 (1968):1364–66.

2. See C. F. Westoff, "Fertility in the United States," *Science* 234 (1986): 554–59.

3. See B. Speck et al., "Allogeneic Bone Marrow Transplantation in a Patient with Aplastic Anemia Using a Phenotypically HL-A Identical Unrelated Donor," *Transplantation* 16 (1974): 24–28.

4. See M. R. Howard et al., "Unrelated Donor Marrow Transplantation between 1977 and 1987 at Four Centers in the United Kingdom," *Transplantation* 49 (1990): 547–53; J. L. Gawejski et al., "Bone Marrow Transplantation Using Unrelated Donors for Patients with Advanced Leukemia or Bone Marrow Failure," *Transplantation* 50 (1990): 244–49; and P. G. Beatty et al., "Marrow Transplantation from HLA-Matched Unrelated Donors for Treatment of Hematologic Malignancies," *Transplantation* 51 (1991): 443–47.

5. See NMDP website, www.marrow.org/NMDP/about_nmdp_idx.html on 6 August 2002.

6. Personal communication, Dennis Confer, Medical Director, NMDP, 1997.

7. Ibid.

8. See T. Nakahata and M. Ogawa, "Hematopoietic Colony-Forming Cells in Umbilical Cord Blood with Extensive Capability to Generate Mono- and Multipotential Hematopoietic Progenitors," *Journal of Clinical Investigation* 70 (1982): 1324–28; and A. G. Leary et al., "Single Cell Origin of Multilineage Colonies in Culture," *Journal of Clinical Investigation* 74 (1984): 2193–97.

9. See H. E. Broxmeyer et al., "Human Umbilical Cord Blood as a Potential Source of Transplantable Hematopoietic Stem/Progenitor Cells," *Proceedings of the National Academy of Sciences* 86 (1989): 3828–32.

10. See M. Ende, "Lymphangiosarcoma: Report of a Case," *Pac Medical Surgery* 74 (1966): 80–82; and M. Ende and N. Ende, "Hematopoietic Transplantation by Means of Fetal (Cord) Blood: A New Method," *Virginia Medical Monthly* 99 (1972): 276–80.

11. See E. Gluckman et al., "Hematopoietic Reconstitution in a Patient with Fanconi's Anemia by Means of Umbilical Cord Blood from an HLA-Identical Sibling," *New England Journal of Medicine* 321 (1989): 1174–78.

12. Personal communication, Pablo Rubinstein, Director, Placental Blood Program, New York Blood Center, 2002.

13. See P. Rubinstein et al., "Outcomes among 562 Recipients of Placental Blood Transplants from Unrelated Donors," *New England Journal of Medicine* 338 (1998):1565–77.

14. See K. A. Jackson et al., "Stem Cells: A Minireview," *Journal of Cellular Biochemistry* 85, issue S38 (2002): 1–6.

15. See E. M. Horwitz et al., "Isolated Allogeneic Bone Marrow–Derived Mesenchymal Cells Engraft and Stimulate Growth in Children with Osteogenesis Imperfecta: Implications for Cell Therapy of Bone," *Proceedings of the National Academy of Sciences* 99 (2002): 8932–37.

16. See E. Gluckman et al., "Outcome of Cord Blood Transplantation from Related and Unrelated Donors," *New England Journal of Medicine* 337 (1997): 373–81.

17. See P. Rubinstein et al., "Processing and Cryopreservation of Placental/Umbilical Cord Blood for Unrelated Bone Marrow Reconstitution," *Proceedings of the National Academy of Sciences* 92 (1995): 10119–22.

18. Personal communication, Pablo Rubinstein, 2002.

19. Personal communication, Dennis Confer, 1997.

20. Personal communication, Pablo Rubinstein, 2002.

21. Personal communication, Dennis Confer, 1997.

22. See P. H. Beatty, M. Mori, and E. Milford, "Impact of Racial Genetic Polymorphism on the Probability of Finding an HLA-Matched Donor," *Transplantation* 60 (1995): 778–83.

23. Personal communication, Dennis Confer, 1997.

24. See J. Sugarman et al., "Ethical Issues in Umbilical Cord Blood Banking," *Journal of the American Medical Association* 278 (1997): 938–43.

25. Personal communication, Pablo Rubinstein, 2002.

26. Ibid.

27. See H. Attarian et al., "Long-Term Cryopreservation of Bone Marrow for Autologous Storage," *Bone Marrow Transplantation* 17 (1996): 425–30; and W. Aird et al., "Long-Term Cryopreservation of Human Stem Cells," *Bone Marrow Transplantation* 9 (1992): 487–90.

28. See Aird et al., "Long-Term Cryopreservation of Human Stem Cells."

29. See Sugarman et al., "Ethical Issues in Umbilical Cord Blood Banking."

30. See ViaCord, "Letter to Participating Physicians," 1997.

CHAPTER
5

Stem Cell Plasticity

Adult Bone Marrow Stromal Cells Differentiate into Neurons

IRA B. BLACK AND
DALE WOODBURY

The past decade has witnessed a remarkable reconceptualization of brain function. The standard model of a complex, though static, system of billions of interconnected nerves (neurons) has been replaced by a model characterized by change (plasticity). Plasticity occurs at multiple levels, from mentation to behavior to systems connectivity to individual neurons to gene activity. A near revolutionary series of recent discoveries has detected new neuron production in the adult brain, long regarded as incapable of regrowth.[1] Emerging evidence indicates that experience can alter the proliferation of new neurons in the brain.[2] In closely related experimental work, stem cells, progenitors capable of developing into multiple cell types, have been found in many organs; and they can be differentiated into neurons and transplanted to the brain to replace diseased and dying cells. In aggregate, these studies are prompting a reformulation of the principles of normal function and are raising the possibility of entirely new approaches to brain disease.

Stem cells have been detected in many adult tissues, participating in normal replacement and repair, while undergoing self-renewal.[3] A subset of bone marrow stem cells is one subtype, capable of differentiating into bone, cartilage, muscle, tendon, fat, and other connective tissue (mesenchymal) derivatives in culture.[4] They have been termed bone marrow stromal cells (BMSCs).

The recent detection of stem cell populations in the central nervous system (CNS) has attracted great attention, since the brain has long been regarded as incapable of regrowth.[5] Neural stem cells (NSCs) undergo expansion and differentiate into neurons and support cells (glia) in vitro.[6] NSCs transplanted into the adult rodent brain survive and differentiate into neurons and glia, raising the possibility of new therapeutic approaches.[7] However, the inaccessibility of NSCs deep in the brain may limit clinical utility. Recently, NSCs were shown to generate blood (hematopoietic) cells in vivo, suggesting that stem cell populations may be less restricted than previously appreciated.[8] Evidence that BMSCs introduced into the lateral ventricles of neonatal mice can differentiate into astrocytes and neurofilament-containing cells—presumptive neurons—lends support to this contention.[9] Increasing evidence now suggests that adult BMSCs can differentiate into presumptive neurons in vitro[10] and in vivo.[11]

These startling new findings are altering our understanding of normal brain function and raising the possibility of novel treatments for now hopeless brain disease. Stem cells and neurons are being transplanted into animal models of Parkinson's, Alzheimer's, and Lou Gehrig's diseases, as well as stroke and spinal cord injury, with the hope of future use in human disorders. Attendant to these opportunities, the discoveries also raise a number of new scientific, clinical, and ethical issues that constitute the background for this brief summary. How do we prevent immunorejection of foreign cells by the host; can we avoid the use of toxic immunosuppressive agents? How do we generate enough cells for each patient to satisfy the potential need for chronic treatment, if necessary? Can we identify safe, accessible sources of cells for long-term therapy? Can we generate cells in a manner that minimizes the risk of tumor formation after transplantation? Which of the many stem cell populations are the most advantageous for clinical use, and how do embryonic, fetal, and adult populations compare? How do we address the ethical concerns associated with the use of embryonic tissue?

To begin approaching these and related issues, we first provide the scientific context, outlining the differentiation of neurons from adult rat and human stromal stem cells. With this background, we begin to address the foregoing issues.

ADULT STROMAL CELLS: CHARACTERIZATION

Rat BMSCs were isolated from the femurs (thigh bones) of adult rats and propagated in vitro.[12] Fluorescent cell sorting at passage one indicated that the cells were negative for CD11b and CD45, surface markers associated with lymphohematopoietic (blood) cells. Consequently, there was no evidence of hematopoietic precursors in the cultures. In contrast, the BMSCs did express CD90, consistent with their undifferentiated state. Initially, with neuronal differentiation, untreated BMSCs were further characterized by staining positively for CD44 and CD71, recapitulating previous reports.[13] These observations are consistent with the contention that this population comprised primitive mesenchymal (mesodermal) cells free of contamination.

ADULT BMSCS DIFFERENTIATE INTO NEURONS

To differentiate BMSCs into neurons, cells were maintained in subconfluent cultures in serum-containing medium supplemented with one mM beta-mercaptoethanol (BME) for twenty-four hours.[14] To induce neuronal differentiation, the cells were transferred to serum-free medium containing 1–10 mM BME (SFM/BME). Within sixty minutes of exposure to SFM/BME, changes in morphology of some of the BMSCs were apparent. Responsive cells progressively assumed neuronal morphological traits over the first three hours. Initially, cytoplasm in the flat BMSCs retracted toward the nucleus, forming a contracted multipolar cell body, leaving membranous, process-like extensions peripherally (0–90 min). Cells exhibited increased expression of the neuronal marker NSE (neuron specific enolase) within thirty minutes of treatment. Over the subsequent two hours, cell bodies became increasingly spherical and refractile, displaying a typical neuronal perikaryal appearance. Processes continued to extend, exhibiting primary and secondary branches, growth cone–like terminal expansions, and putative filopodial extensions, characteristic of neurons.

To further characterize neuronal differentiation, BME-treated cultures were fixed after five hours and stained for the neuronal marker NSE.[15] Unresponsive, flat BMSCs expressed very low, but detectable, levels of NSE protein, consistent with previous detection of minute amounts of protein and/or message in cells of bone marrow origin.[16] Progressive transition of BMSCs to a neuronal phenotype coincided with increased expression of NSE. BMSC-derived neurons displayed distinct neuronal morphologies, ranging from simple bipolar to large, extensively branched multipolar cells. Clusters of differentiated cells exhibited intense NSE positivity, and processes formed extensive networks. Western blot analysis confirmed the expression of low levels of NSE protein in uninduced BMSCs. Induction of the neuronal phenotype dramatically increased NSE expression, consistent with the immunocytochemical data.

To further characterize the putative neurons, BMSCs were treated with DMSO/BHA (dimethylsulfoxide and butylated hydroxyanisole, an antioxidant) and were stained for neurofilament-M (NF-M), a neuron-specific intermediate filament that helps initiate neurite elongation.[17] We had previously found that BME treatment of BMSCs increased expression of NF-M in cells exhibiting neuronal morphologies. Most cells displaying rounded cell bodies with processes after DMSO/BHA exposure expressed high levels of NF-M, whereas flat, undifferentiated cells did not. Preadsorption of NF-M antibody with purified NF-M protein abolished staining, establishing specificity.[18]

We examined DMSO/BHA-treated cultures for the presence of tau, a neuron-specific microtubule-associated protein expressed by differentiating neurons.[19] Cells exhibiting neuronal morphologic traits expressed tau protein in the cell body as well as in the processes, whereas undifferentiated flat cells were tau-negative.[20]

To investigate neuronal traits further, we examined differentiated cultures for NeuN, a neuron-specific marker expressed in postmitotic cells.[21] A subset of cells exhibiting rounded cell bodies and processes stained for NeuN expression, whereas neighboring cells exhibiting distinct neuronal morphologies were NeuN-negative. This observation suggests that subsets of NSE-positive cells are postmitotic neurons.

QUANTITATIVE ANALYSIS

Using this induction protocol, a variable number of cells underwent neuronal differentiation, though the response generally exceeded 50 percent. To optimize differentiation, we modified the preinduction protocol. Addition of bFGF (basic fibroblast growth factor; 10 ng/ml) and elimination of BME from the preinduction medium increased the proportion of cells exhibiting neuronal traits, and the response was more consistent within and between experiments. To quantitate this response, BMSCs were treated with the modified preinduction medium and induced to differentiate with DMSO/BHA. Cells were fixed after five hours, stained for the neuronal markers NSE and NF-M, and the percentage of presumptive neurons was defined. The majority of BMSCs treated in this manner exhibited neuronal morphologies and were immunopositive for NSE (78.2% ± 2.3%) and NF-M (79.2% ± 2.5%).[22]

LONG-TERM NEURONAL DIFFERENTIATION

The rapidity of BMSC neuronal differentiation is one striking characteristic of the response. Consequently, we initially focused on changes that occurred within the first five hours of differentiation. However, long-term differentiation of these cells will be important to our understanding of this process. To address this issue, we monitored expression of the nestin gene product in BMSC-derived neurons at five hours, one day and six days postdifferentiation. Nestin, an intermediate filament protein, is normally expressed in neuroepithelial neuronal precursor stem cells, and its expression decreases with neuronal maturation.[23] A subset of BMSC-derived neurons expressed high levels of nestin protein at five hours, and the proportion of nestin-positive cells decreased with time. By six days postinduction, there was no detectable nestin expression in any BMSC-derived neurons, consistent with ongoing maturation with time.[24]

We compared the time course of expression of the transitional trait, nestin, to that of a mature phenotypic character, trkA, the biologically active nerve growth factor receptor. TrkA was detectable at five hours, the earliest time examined and, in contrast to nestin, persisted unchanged through six days.[25] In sum, these observations suggest that neuronal differentiation of adult BMSCs follows the normal developmental sequence exhibited by neurons in vivo.

ANALYSIS AT THE CLONAL LEVEL

To determine whether individual BMSCs exhibited stem cell characteristics of self-renewal and pluripotentiality, individual clones were analyzed. To establish clones, BMSCs were plated at approximately 10 cells/cm^2, grown to 50–150 cells per colony, isolated with cloning cylinders, and transferred to separate wells and finally to individual flasks. Single cells replicated as typical BMSCs and differentiated into NSE-positive neurons after BME treatment. In fact, each individual clone generated refractile, process-bearing, NSE-positive cells following BME treatment. Undifferentiated BMSCs and transitional cells were evident in each clonal line. Consequently, clones derived from single cells gave rise to both BMSCs and neurons, indicating stem cell characteristics.[26]

ADULT HUMAN STROMAL CELLS DIFFERENTIATE INTO NEURONS

To determine whether the neuronal potential of BMSCs is unique to rodents, or whether human BMSCs (hBMSCs) share this ability, BMSCs were isolated from a healthy adult donor and grown in vitro.[27] Cells from passage two were subjected to the neuronal differentiation protocol and examined for NSE or NF-M expression. After BME treatment, hBMSCs exhibited neuronal characteristics and increased NSE expression in a time frame similar to that observed for rat BMSCs. Contracted cell bodies elaborated processes and stained intensely for NSE expression within three hours. Transitional cells were also evident. Many processes elaborated by hBMSC-derived neurons exhibited terminal bulbs, which may represent growth cones. These cells also expressed NF-M, consistent with neuronal differentiation.[28]

FUTURE DIRECTIONS AND PERSPECTIVE

These early studies suggest that traditional models of development and differentiation may require reexamination. Intrinsic genomic mechanisms of commitment, lineage restriction, and cell fate may be more flexible than had been assumed. In our investigations, adult BMSCs, derivatives of the embryonic mesoderm, differentiate into presumptive neurons, classical ectodermal cells. Environmental signals apparently elicit the expression of pluripotentiality that extends well beyond the conventional fate restrictions of cells derived from the classical embryonic germ layers. In addition to potentially altering views of differentiative mechanisms and potentials, these observations may inform new therapeutic approaches. We have transplanted the BMSCs and neurons into the brains and spinal cords of normal and disordered rats and are presently analyzing results. While the use of BMSCs may confer a number of advantages, a host of open questions emphasizes the rudimentary nature of our knowledge.

Some of the theoretical advantages of using adult BMSCs may be delineated as background, prior to articulating questions and problems. Use of the patient's own bone marrow cells (autologous transplantation) eliminates the hazard of immunorejection and obviates the need for toxic immunosuppressive agents. Moreover, the bone marrow is a safe, convenient source of tissue, accessible through a routine, bedside procedure. The BMSCs grow vigorously in culture, eliminating the need for immortalization, allowing expansion of the population, and providing a vast reservoir of tissue. The use of a simple in vitro protocol to differentiate neurons, in the absence of genetic manipulation, may reduce the probability of tumor formation after transplantation. In one final example, the use of adult cells circumvents the concerns attendant to the use of embryonic and fetal tissue. However, the last point requires extensive qualification.

Embryonic stem cells remain the gold standard in terms of potentiality and plasticity. It is critical to compare the differentiation of embryonic and adult cells in culture and in vivo to identify the optimal population for any specific application. How do embryonic, fetal, and adult cells compare with respect to population homogeneity, growth in culture, longevity and renewal, multipotentiality, manipulability, maintenance of the appropriate phenotype, biologic functionality, and propensity for tumor formation, to cite only several issues? Upon transplantation, how do the different cells compare regarding homing to the correct site, engraftment, survival, functional integration into the host tissue, and recovery of function? If cells are to be used for gene therapy, which populations are most amenable to the insertion (transfection) of new genes and prolonged gene expression in vivo? Which populations are maximally responsive to growth factors that may be employed in combined therapy? How do putative stem cells from different organs at different ages compare regarding these different criteria? In addition to optimizing clinical efficacy, toxicity of different cells must be assessed. How do embryonic, fetal, and adult cells compare regarding immunogenicity, inflammatory potential, secretion of potentially toxic products, and tumorgenicity? Comparative studies are just beginning to address these critical issues. Until we have answers to these pressing questions, it will be difficult to formulate optimal therapeutic approaches that maximize efficacy and minimize side effects and toxicity. We are presented with a remarkable opportunity that now mandates meticulous, systematic approaches to help reformulate mechanisms of development and treatment of disease.

NOTES

The National Institutes of Health grant HD23315, the New Jersey Commission for Spinal Cord Research, the Christopher Reeve Paralysis Foundation, and the Michael J. Fox Foundation supported this work.

1. See B. A. Reynolds and S. Weiss, "Generation of Neurons and Astrocytes from Isolated Cells of the Adult Mammalian Central Nervous System," *Science* 255 (1992): 1707–10; H. A. Cameron et al., "Differentiation of Newly Born Neurons and Glia in the Dentate Gyrus of the Adult Rat," *Neuroscience* 56 (1993): 337–44; and F. H. Gage, "Neurogenesis in the Adult Brain," *Journal of Neuroscience* 22, no.3 (2002): 612–13.

2. See H. van Praag et al., "Running Increases Cell Proliferation and Neurogenesis in the Adult Mouse Dentate Gyrus," *Nature Neuroscience* 2 (1999): 226–70; E. Gould et al., "Learning Enhances Adult Neurogenesis in the Hippocampal Formation," *Nature Neuroscience* 2 (1999): 260–65; and E. Gould et al., "Proliferation of Granule Cell Precursors in the Dentate Gyrus of Adult Monkeys Is Diminished by Stress," *Proceedings of the National Academy of Sciences* 95 (1998): 3168–71.

3. See A.I. Caplan, "Mesenchymal Stem Cells," *Journal of Orthopaedic Research* 9 (1991): 641–50; G. Ferrari et al., "Muscle Regeneration by Bone Marrow–Derived Myogenic Progenitors," *Science* 279 (1998): 1528–30; E. Hay, *Regeneration* (New York: Holt, Rinehart and Winston, 1966); S.A. Kuznetsov et al., "Factors Required for Bone Marrow Stromal Fibroblast Colony Formation in Vitro," *British Journal of Haemotology* 97 (1997): 561–70; I. Lemischka, "Searching for Stem Cell Regulatory Molecules. Some General Thoughts and Possible Approaches," *Annals of the New York Academy of Sciences* 872 (1999): 274–88; M.K. Majumdar et al., "Phenotypic and Functional Comparison of Cultures of Marrow-Derived Mesenchymal Stem Cells (MSCs) and Stromal Cells," *Journal of Cell Physiology* 176 (1998): 57–66; R.D. McKay, "Brain Stem Cells Change Their Identity," *Nature Medicine* 5 (1999): 261–62; M.E. Owens and A.J. Friedenstein, "Cell and Molecular Biology of Vertebrate Hard Tissues," *Ciba Foundation Symposium* 136 (1988): 42–60; R.F. Pereira et al., "Cultured Adherent Cells from Marrow Can Serve as Long-Lasting Precursor Cells for Bone, Cartilage, and Lung in Irradiated Mice," *Proceedings of the National Academy of Sciences* 92, no. 11 (1995): 4857–61; M.F. Pittenger et al., "Multilineage Potential of Adult Human Mesenchymal Stem Cells," *Science* 284 (1999): 143–47; D.J. Prockop, "Marrow Stromal Cells as Stem Cells for Nonhematopoietic Tissues," *Science* 276 (1997): 71–74.

4. See Caplan, "Mesenchymal Stem Cells"; Ferrari et al., "Muscle Regeneration by Bone Marrow–Derived Myogenic Progenitors"; Kuznetsov et al., "Factors Required for Bone Marrow Stromal Fibroblast Colony Formation in Vitro"; Majumdar et al., "Phenotypic and Functional Comparison of Cultures of Marrow-Derived . . . (MSCs) . . .; Pereira et al., "Cultured Adherent Cells from Marrow . . . in Irradiated Mice"; Pittenger et al., "Multilineage Potential of Adult Human Mensenchymal Stem Cells"; and Prockop, "Marrow Stromal Cells as Stem Cells. . . ."

5. See Reynolds and Weiss, "Generation of Neurons and Astrocytes . . ."; and L.J. Richards et al., "De Novo Generation of Neuronal Cells from the Adult Mouse Brain," *Proceedings of the National Academy of Sciences* 89: 8591–95.

6. See F.H. Gage et al., "Survival and Differentiation of Adult Neuronal Progenitor Cells Transplanted to the Adult Brain," *Proceedings of the National Academy of Sciences* 92 (1995): 11879–83; C.B. Johansson et al.,

"Identification of a Neural Stem Cell in the Adult Mammalian Central Nervous System," *Cell* 96 (1999): 25–34; Reynolds and Weiss "Generation of Neurons and Astrocytes . . ."; and A. L. Vescovi et al., "bFGF Regulates the Proliferative Fate of Unipotent (Neuronal) and Bipotent (Neuronal/Astroglial) EGF-Generated CNS Progenitor Cells," *Neuron* 11 (1993): 951–66.

7. See J.D. Flax et al., "Engraftable Human Neural Stem Cells Respond to Developmental Cues, Replace Neurons, and Express Foreign Genes," *Nature Biotechnology* 16 (1998):1033–39; F. H. Gage et al., "Isolation, Characterization, and Use of Stem Cells from the CNS," *Annual Review of Neuroscience* 18 (1995): 159–92; C. Lundberg et al., "Survival, Integration, and Differentiation of Neural Stem Cell Lines after Transplantation to the Adult Striatum," *Experimental Neurology* 145 (1997): 342–60; P. J. Renfranz et al., "Region-Specific Differentiation of Hippocampal Stem Cell Line HiB5 upon Implantation into the Developing Mammalian Brain," *Cell* 6 (1991): 713–29; and C. N. Svendsen et al., "Long-Term Survival of Human Central Nervous System Progenitor Cells Transplanted into a Rat Model of Parkinson's Disease," *Experimental Neurology* 148 (1997): 135–46.

8. See C. R. Bjornson et al., "Turning Brain into Blood: A Hematopoietic Fate Adopted by Adult Neural Stem Cells in Vivo," *Science* 283 (1999): 534–37.

9. See G. C. Kopen et al., "Marrow Stromal Cells Migrate throughout the Forebrain and Cerebellum, and They Differentiate into Astrocytes after Injection into Neonatal Mouse Brains," *Proceedings of the National Academy of Sciences* 96 (1999): 10711–16.

10. See H. B. Sarnat et al., "Neuronal Nuclear Antigen (NeuN): A Marker of Neuronal Maturation in Early Human Fetal Nervous System," *Brain Research* 20 (1998): 88–94; and D. Woodbury et al., "Adult Rat and Human Bone Marrow Stromal Cells Differentiate into Neurons," *Journal of Neuroscience Research* 61 (2000): 364–70.

11. See T. R. Brazelton et al., "From Marrow to Brain: Expression of Neuronal Phenotypes in Adult Mice," *Science* 290 (2000): 1775–79; and E. Mezey et al., "Turning Blood into Brain: Cells Bearing Neuronal Antigens Generated in Vivo from Bone Marrow," *Science* 290 (2000): 1779–82.

12. See S. A. Azizi et al., "Engraftment and Migration of Human Bone Marrow Stromal Cells Implanted in the Brains of Albino Rats—Similarities to Astrocyte Grafts," *Proceedings of the National Academy of Sciences* 95 (1998): 3908–13.

13. See S. P. Bruder et al., "Mesenchymal Stem Cells in Osteobiology and Applied Bone Regeneration," *Clinical Orthopaedics and Related Research* 355S (1998): S247–S256; and Pittenger et al., "Multilineage Potential of Adult Human Mesenchymal Stem Cells."

14. See Woodbury et al., "Adult Rat and Human Bone Marrow Stromal Cells Differentiate into Neurons."

15. See ibid.

16. See H. Pechumer et al., "Detection of Neuron Specific Enolase Messenger Ribonucleic Acid in Normal Human Leukocytes by Polymerase Chain Reaction Amplification with Nested Primers," *Laboratory Investigation* 89 (1993): 743–49; M. M. Reid et al., "Routine Histological Compared with Immunohistological Examination of Bone Marrow Trephine Biopsy Specimens in Disseminated Neuroblastoma," *Clinical Pathology* 44 (1991): 483–86; and E. van Obberghen et al., "Human γ-Enolase: Isolation of a cDNA Clone and Expression in Normal and Tumor Tissues of Human Origin," *Journal of Neuroscience Research* 19 (1988): 450–56.

17. See M. J. Carden et al., "Two-Stage Expression of Neurofilament Polypeptides during Rat Neurogenesis with Early Establishment of Adult Phosphorylation Patterns," *Neuroscience* 7 (1987): 337–44.

18. See Woodbury et al., "Adult Rat and Human Bone Marrow Stromal Cells Differentiate into Neurons."

19. See K. S. Kosik et al., "MAP2 and Tau Segregate into Dendritic and Axonal Domains after the Elaboration of Morphologically Distinct Neurites: An Immunocytochemical Study of Cultured Rat Cerebrum," *Journal of Neuroscience* 7 (1987): 3142–53.

20. See Woodbury et al., "Adult Rat and Human Bone Marrow Stromal Cells Differentiate into Neurons."

21. See H. B. Sarnat et al., "Neuronal Nuclear Antigen (NeuN): A Marker of Neuronal Maturation in Early Human Fetal Nervous System," *Brain Research* 20 (1998): 88–94.

22. See Woodbury et al., "Adult Rat and Human Bone Marrow Stromal Cells Differentiate into Neurons."

23. See U. Lendahl et al., "CNS Stem Cells Express a New Class of Intermediate Filament Protein," *Cell* 60 (1990): 295–300.

24. See Woodbury et al., "Adult Rat and Human Bone Marrow Stromal Cells Differentiate into Neurons."

25. See ibid.

26. Ibid.

27. Ibid.

28. Ibid.

PART II

Ethical Issues in Stem Cell Research

The complexity of the scientific and public policy questions prompted by stem cell research is matched, if not exceeded, by that of the ethical issues this research raises. The ethical dimensions of stem cell research are vigorously examined in the essays that comprise part II of this volume.

In "Stem Cell Ethics: Lessons from the Context," Karen Lebacqz argues for the moral permissibility of stem cell research on the basis of the principle of respect for persons. Central for her argument are naming the ethical issues, taking context seriously, and formulating the questions correctly. For stem cell research, the two central ethical issues are whether it is morally permissible to destroy blastocysts and whether it is permissible to create blastocysts for research purposes. Lebacqz maintains that strong moral convictions about prohibiting the deliberate destruction of innocent persons as well as prohibiting the use of persons as means to others' ends raise troubling issues for stem cell research.

She begins by examining the practice of in vitro fertilization (IVF). She argues that if IVF is accepted, as seems to be the case, then stem cell research involving the creation and destruction of embryos should also be accepted. IVF as currently practiced, she argues, shows widespread public acceptance of the destruction of embryos and of the deliberate creation of embryos to be used as means to another's ends. Lebacqz can find no morally relevant differences between the purposes and contexts of IVF and stem cell research that would justify accepting the creation and destruction of embryos in one context but not in the other.

She stresses that she is no proponent of IVF. Her point is that if IVF is accepted, the ethical issues at stake in stem cell research need to be reformulated to ensure consistency of argumentation across both contexts. Debates about the ethics of stem cell research usually center on whether the embryo is a person or not. This structure, Lebacqz maintains, is flawed. It can be wrong to kill a redwood tree even though it is not a person, yet permissible to kill a person. We generally think that it is not respectful to deliberately kill a person because deliberate killing violates autonomy, causes pain or suffering, and cuts off the person's destiny. However, there are circumstances in which deliberate killing is respectful. Lebacqz argues that when there is no autonomy, no sentience (and thus, no pain or suffering is caused by killing), and the person's destiny is death under any circumstances, direct killing is not wrong, for it is not disrespectful. She thus concludes that blastocysts that are slated for destruction anyway may be destroyed for research purposes.

What about the creation of embryos for research purposes? She proposes a regulative principle: all human embryos that are to develop should be created only by means that can reasonably be expected to present no harms to the embryo beyond those of ordinary development in the womb. In stem cell research, the fact that embryos will not develop furnishes protection against harming a child-to-be. This protection is not provided in IVF or reproductive cloning. Thus, on this principle, it is worse to permit IVF than stem cell research.

Lebacqz points out that her arguments are not utilitarian, but are based on principle. According to her, the fundamental principle at stake in the stem cell research debate is not the prohibition against killing, but that of respect for persons. What does respect for persons require? Where there is no autonomy and no sentience, she concludes, it may not require an absolute prohibition against killing or against using another as a means only. Consequently, stem cell research does not violate the principle of respect for persons.

May a Catholic researcher or scientist receive embryonic stem cells after the human embryo has been destroyed in order to establish an embryonic stem cell line? May a Catholic researcher or scientist use an established embryonic stem cell line in order to discover new drugs or therapies to cure debilitating disease? May a Catholic physician or patient use a new drug or therapy that has been derived using an embryonic stem cell line for the cure or treatment of a disease? Using the principles of cooperation found in Catholic moral theology, Edward J. Furton analyzes these questions in "Levels of Moral Complicity in the Act of Human Embryo Destruction."

Furton starts with the premise that human embryo destruction is intrinsically evil. He proceeds to define "cooperation" as help afforded another, who is not seduced, to carry out the purposes of sinning. He then distinguishes among explicit and implicit formal cooperation, proximate and remote material cooperation, and mediate and immediate material cooperation. Formal cooperation occurs when one shares the evildoer's intention. It is explicit when one intends the same immoral act as the wrongdoer. It is implicit when one does not share the same intention, but is willing to participate in the same activity, albeit for an ostensibly different purpose than that of the wrongdoer. Material cooperation occurs when one aids the external commission of a sin, but does not intend the sin. Depending upon the nearness and the definiteness of the assistance provided, it can be either proximate or remote. Material cooperation can be blameless if there is a sufficiently weighty reason for it, but without such a reason, the action is immoral. The most controversial of these distinctions is that between immediate and mediate material cooperation. Furton endorses the view that material cooperation is immediate or mediate depending upon one's sharing in the sinful act of the wrongdoer, or in some act that preceded or followed it. He advises that his analysis depends upon seeing cooperative action as a causal contribution to evil. If the cooperator adds no causal dimension to a wrongful action, he claims, then the cooperator's contribution is either above moral reproach or is not cooperation at all.

The application of the principles of cooperation to human embryonic stem cell research is facilitated by considering similarities and differences of a parallel case of vaccines that were developed from distant cases of abortion. Drawing on the principles and the vaccine case, Furton concludes that a person who destroys human embryos commits an act of wrongdoing. One who assists by sharing in the intention of this act would be engaged in formal cooperation. One who receives the inner cells derived from the destroyed embryo receives the remains of a human being. If the recipient does not share the intention of the person who destroys the embryo, the act of receiving the cells does not constitute formal cooperation. Furton believes, however, that the act of receiving the cells can be subsumed under the heading of immediate material cooperation. The coordinated activities of receiving and handling these cells makes the recipient party to the original act of destruction.

What about working with established stem cell lines? These are the daughter cells of the embryo, not the original cells derived from its destruction. Can a Catholic researcher or scientist blamelessly use such established cell lines to develop new drugs or therapies? Furton's answer is complex. Though Catholic researchers who work with established cell lines do not directly participate in an intrinsically evil act, they become part of a larger research community that includes others who are engaged in intrinsically evil acts of embryo destruction. Moreover, Furton contends, once research on a stem cell line has begun, it is very difficult to change course. Scientific progress that is made using these stem cell lines could lead people to believe that it is good to destroy human embryos. At this stage, Furton believes that we have the opportunity to begin stem cell research aright. Consequently, he concludes that Catholic researchers engage in mediate material cooperation by using embryonic stem cell lines and that they have an obligation to forego such research.

Furton's final question is whether Catholic physicians and patients may use drugs or therapies produced by research on embryonic stem cell lines. He argues that such use would constitute remote material cooperation. At this level of cooperative distance, however, it is morally permissible to benefit from the products of embryonic stem cell research.

In "Stem Cells and Social Ethics: Some Catholic Contributions," Lisa Sowle Cahill identifies five general values that should be protected in the debate about human embryonic stem cell research: the value of nascent life, moral virtue or moral integrity, the value of medical benefits, distributive justice or just institutions, and the value of a social ethos of generosity or solidarity. She discusses the values using five principles of Catholic theology: probabilism, cooperation, double effect, common good, and the preferential option for the poor. The position she constructs is a "convergence argument"—an ethical outlook based on several arguments. Though none of these arguments is conclusive by itself, the position that Cahill forges suggests a general perspective on the ethics of stem cell research. Central to this perspective, she thinks, is advocacy for basic health care and drawing the firmest line possible on creating or cloning embryos for research purposes.

In her discussion of the value of nascent life, Cahill acknowledges that the moral status of the embryo is highly contentious. She reviews several positions held by Roman Catholic moral theologians and secular ethicists. Given that there is doubt or ambiguity about the moral status of the embryo, Cahill invokes the doctrine of probabilism. According to this approach, the greater the doubt about the rectitude of a moral teaching, the more flexibility there can be in the application of the teaching. To what extent does doubt about the moral status of the embryo allow for flexibility in applying the principle of the inviolability of human life? Taking a probabilistic approach to this question requires weighing numerous factors, such as the value of medical benefits and the values of distributive justice and solidarity—taken up by Cahill later in her article.

Cahill's discussion of the value of moral virtue centers on moral integrity. Moral integrity arises in the stem cell research debate in terms of cooperation with or complicity in the evil (not yet determined to be a moral evil) of destroying embryos. Moral concerns raised by the term "complicity" fall into two categories: those affecting the moral standing of the agent who proposes to be associated with the evil actions of another; and the effects of complicity on the social context or community within which the actions occur. Cooperation identifies similar but not identical concerns. Like the principle against complicity, the principle of cooperation addresses the intention and moral position of the agent, as well as the agent's role as a possible facilitator of evil. One difference, according to Cahill, is that cooperation judges actions prospectively, whereas complicity looks back to past actions whose consequences one now appropriates as part of one's own action. Despite an extensive and complex discussion of the principles of cooperation and complicity, Cahill concludes that, by themselves, they are not decisive of the ethics of stem cell research.

What role should the value of medical benefits play in the ethics of stem cell research? To be more precise, are we justified in causing an evil (the death of embryos) in pursuit of the potential medical benefits to be provided by stem cell therapies? Cahill deploys the principle of double effect to address this question. The principle permits the performance of an action that produces both a good and an evil effect if four conditions are met: (1) the act, considered in itself independently of its effects, must not be morally evil; (2) the evil effect must not be the means of producing the good effect; (3) the evil effect is not intended but merely tolerated; and (4) there must be a proportionate reason for performing the act, despite its evil consequences. Applying the doctrine to the stem cell research debate leads us back to two central but unresolved questions. First, what is the moral status of the embryo? That is, is the death of the embryo a moral evil, and if so, of what magnitude? Second, how great are the medical benefits to be gained from stem cell research? Lacking clear answers to these questions, the results of applying the doctrine of double effect remain ambiguous. Despite this, the doctrine is a useful reminder, according to Cahill, of the fact that the principle of utility ("the greatest good for the greatest number") is not self-sufficient.

Cahill's last two topics, distributive justice and the value of a social ethos of generosity and solidarity, urge us to take a broader perspective on the stem cell research debate. Distributive justice issues are part of the Catholic common good tradition, and they concern fairness in the distribution of benefits and burdens, as well as the justice, of social institutions. Since the 1960s, Catholic social teaching on the common good has included a global dimension. Consequently, we should be concerned, Cahill thinks, with who has access to the potential benefits of stem cell therapies. She points out that in our nation of forty-four million uninsured people, the group of beneficiaries is not likely to be inclusive. Moreover, stem cell therapies will be exotic as well as expensive to millions of the world's population, who are beset by poverty and disease and who suffer from the lack of food, clean drinking water, and basic health care. Finally, she indicates that, if therapies are developed from stem cells derived from embryos left over from IVF, they are unlikely to provide good tissue matches for a diversity of populations.

What kind of society do we want to create? Cahill argues for a social ethos of generosity and solidarity, informed by the principle of a preferential option for the poor. This ethic runs counter to current social trends of commercialization and commodification. Cahill reminds us that when basic human goods are commodified, some people who need them are likely to be excluded. Further, destroying embryos and selling their parts increases commodification in the area of procreation, which should, she thinks, be an expression of sexual, parental, and familial commitments. She believes that generosity and solidarity, as opposed to commodification and the profit motive, are the virtues that we should strive for in struggling with questions of the treatment of the embryo and the distribution of lifesaving therapies. Taking this approach does not mean that the biotechnology, medical, or health care markets must be eliminated. It does require that they be framed and circumscribed by larger, more inclusive values. Similarly, debates about the ethics of stem cell research should be framed within the context of larger concerns for social justice and equal human dignity and respect.

Another kind of convergence argument is constructed by Richard M. Doerflinger in "The Ethics and Policy of Embryonic Stem Cell Research: A Catholic Perspective." Doerflinger identifies six propositions in the stem cell research debate on which there is broad, though not unanimous, consensus. These propositions, he claims, are not specific to any religious perspective and provide the basis for a strong argument against proceeding with human embryonic stem cell research. (1) At the blastocyst stage, the human embryo is a developing human life. Doerflinger argues that one need not claim that the embryo has the moral status of a person in order to understand the moral obligation to respect and protect developing human life. (2) Pursuing medical treatments through human embryonic stem cell research requires us to condone the destruction of this developing human life. According to Doerflinger, human embryonic stem cell research implies this attitude toward developing human life. He discusses several arguments and policies that obscure this fact, including President Bush's compromise policy of allowing federally funded research on stem cell lines derived from human embryos before August 9, 2001. (3) A moral presumption against taking human life requires us to treat embryonic stem cell research as a last resort. (4) Adult stem cell research and other alternatives are more promising than once thought, and they offer many of the benefits once thought achievable only with human embryonic stem cells. (5) There are more drawbacks presented by human embryonic stem cell research than once thought. (6) Propositions four and five indicate that the potential clinical benefits of human embryonic stem cell research can be obtained in less problematic ways. Based on these propositions and the arguments for them, Doerflinger concludes, using tax dollars to fund human embryonic stem cell research would be irresponsible.

What, then, is the Catholic contribution to the debate? Doerflinger argues that it consists in identifying alternative assumptions to those held by many scientists and advisory bodies. The latter, he thinks, embrace materialist assumptions about human nature, utilitarian moral reasoning, and the technological imperative that if something is possible, it should be done. Against this worldview, he offers five elements of the Catholic perspective: a faith-based metaphysical realism, according to which there is an objective reality to be discovered; the belief that humans neither create nor control the world; the belief that choices in matters of life and death have eternal consequences; endorsement of a preferential option for the poor and marginalized in matters of the common good; and a concern to defend rights of conscience. This perspective should influence not only the stem cell research debate, but also the related cloning controversy. The convergence between Catholic thought and the views of many others in these debates, Doerflinger urges, is not owing to a "Catholic takeover," but to the fact that Catholic reflections are not as far from "mainstream" thought as some assume.

M. Therese Lysaught begins her contribution, "'What Would You Do If . . . ?' Human Embryonic Stem Cell Research and the Defense of the Innocent," by reflecting on the traditional Catholic commitment to healing and health care. Catholic arguments about stem cell research, she urges, should be viewed within this larger context. Lysaught maintains that the rhetoric of much current discussion about biotechnology, including the stem cell research debate, is pervaded by the language and images of war and aggression. In such circumstances, innocent life can sometimes legitimately be sacrificed. Central to the Christian tradition are three classic examples of cases in which innocent life may legitimately be sacrificed: St. Thomas Aquinas's justification of self-defense; the defense of one's family or neighbor against a malicious attacker; and the just war tradition. Each example, Lysaught contends, shares structural affinities with the stem cell research debate. First, in each situation, the life of an "innocent" has been or is being attacked. Second, in each situation, taking human life is presented as the only, primary, or last option and is required to "defend" the innocent party. She offers a close analysis of each case. Despite some similarities with human embryonic stem cell research, she concludes that none of the examples supplies a sufficiently close parallel with stem cell research to warrant the destruction of embryos this research requires.

For example, she analyzes putative structural affinities between the conditions that justify an act of killing in self-defense for Aquinas and the act of destroying human embryos for the sake of stem cell research. Both acts could be performed with the requisite intention—the intention to save life—and by the requisite person—a public authority. Yet closer scrutiny reveals that these structural similarities are merely superficial. Lysaught's careful scrutiny shows that Aquinas's framework, which seeks to minimize violence we might inflict on one another in the name of our own needs, desires, and justice, does not supply moral conditions that justify destroying human embryos for stem cell research.

The second framework Lysaught examines is the attempted justification of killing an assailant in order to defend an innocent third party. The framework is presented in the question: "What would you do if . . . ?" Might this approach justify destroying embryos for stem cell research? Would we be justified in condoning human embryonic stem cell research for the sake of saving loved ones? Lysaught draws on John Howard Yoder's approach to the "what if?" question, which has been used to challenge pacifism. She unpacks the assumptions and dynamics operative in the stem cell research debate. She finds in the rhetoric of the debate deterministic assumptions about how the science will develop; the presumption of significant control over scientific outcomes; the assumption that we now have or will have accurate and complete knowledge of the behavior of the scientific mechanisms at work in cell differentiation and tissue development; and the presupposition that we do not have promising alternatives to human embryonic stem cell research and the potential benefits it might yield. She concludes that the crux of the "what if" question and the case for human embryonic stem cell research rely mainly on emotional appeal rather than rational argument.

The third framework to be considered is the just war tradition. Lysaught admits that, of the three analogies, this tradition offers the closest fit with the situation of human embryonic stem cell research. One could admit that humanity has a right to defend itself against disease. Despite this, she argues that two important criteria of the just war tradition—the principles of last resort and noncombatant immunity—are not met. Alternatives to human embryonic stem cell research, such as adult stem cell research, have not been exhausted, so the principle of last resort is not satisfied. Further, frozen embryos occupy the role of noncombatants. A larger point, however, is this: As with Aquinas's discussion of the conditions under which killing in self-defense is justifiable, the just war tradition provides a moral framework that seeks to limit the conditions under which violence can legitimately be inflicted. It does not seek to carve out a space for the legitimacy of violence.

As long as human embryos qualify as human life, "sacrificing" them, Lysaught maintains, is not a moral option. Reflecting that the language in which we frame our arguments mirrors the moral world we inhabit, she reminds us that Christians understand healing in relation to peace, not war.

CHAPTER 6

Stem Cell Ethics

Lessons from the Context

KAREN LEBACQZ

A crucial issue in stem cell research is the status of the blastocyst, or "early embryo." I will argue that the widespread acceptance of the practice of in vitro fertilization (IVF) lays the foundation for acceptance of stem cell research and that, if stem cell research is to be rejected, the practice of IVF must also be rejected.[1] If IVF is accepted, then the ethical issues need to be reformulated, as the arguments typically made on both sides do not hold consistently. A careful look at typical arguments further suggests that they are flawed and that there is room for moral approbation of destruction of human life under some circumstances. Those circumstances may include stem cell research.

NAMING THE ETHICAL ISSUES

Two issues dominate debate about the status of the blastocyst from which stem cells are derived. First, is deliberate destruction of a blastocyst permissible? The human embryonic stem cell research that has thus far proved most promising involves the destruction of an embryo at the blastocyst or "hollow ball" stage. The National Bioethics Advisory Commission (NBAC) proposed that the destruction of blastocysts in embryonic stem cell research could be justified under certain conditions. This is also the position taken by the Geron Ethics Advisory Board. However, the deliberations of these bodies have not satisfied many people—and, indeed, have sparked some recent criticism of national ethics boards.

For many people, the deliberate destruction of embryos has already been settled by the legal decision to permit abortion. In a legal climate that permits a woman to have an abortion at will, any time up to the end of the second trimester, it may seem strange to ask whether an embryo at the very earliest stages can be killed. If a late-stage fetus can be killed, why not an early embryo? However, the legal situation does not fully satisfy our convictions about what should be morally permissible. The reason is simple: killing people is morally prohibited under most circumstances. If the blastocyst is a person, morally speaking, then destruction of the blastocyst is prohibited by the general injunction against killing people. Hence, for many people, the destruction of the blastocyst alone renders stem cell research morally impermissible. This is the view held by many Roman Catholics, and it is a view that has had significant influence on public policy. For example, President Bush's compromise position refused to permit the use of federal funds in research that directly destroys the blastocyst. Bush here follows earlier law prohibiting the use of federal funds in research that destroys embryos. Bush's compromise will not satisfy everyone, however; for many believe that even the use of existing stem cell lines makes researchers complicit in the original destruction of the embryos. Both direct and "indirect" destruction of the blastocyst are seen by many people as wrongful taking of human life.

A second ethical issue also arises around this technology: Is creation of a blastocyst for research purposes permissible? This issue arises largely because of the need to overcome problems of histocompatibility if stem cells are ever to be useful for treatments of disease. Stem cells and tissues derived from them will carry the genetic characteristics of the embryo from which the stem cells were derived. While there is some evidence that tissue rejection may not be as strong an issue in stem cell implantation as it has been in organ transplantation, any attempt to implant cells or tissues into a living human being runs some risks of tissue rejection. To avoid incompatibility, one approach would be to use somatic cell nuclear transfer to create an embryo whose stem cells would be genetically compatible with the patient. Another approach would be to develop "universal donor" lines of stem cells. A third would be to use somatic cell nuclear transfer for basic research geared to learning how any cell with differentiated DNA might be triggered to behave as though it is undifferentiated, in hopes that the body can eventually be triggered to heal itself. All three of these approaches require, in the immediate future, the deliberate creation of embryos for research purposes.

The Human Embryo Research Panel did approve the creation of embryos for research purposes under limited circumstances, which would apply to stem cell research. However, the work of this panel has been criticized in this precise arena, and the deliberations of this public body have not settled the moral issues for many. Rather, the general consensus has not supported the deliberate creation of human embryos for research purposes.

Western tradition retains a strong moral mandate to treat others as ends in themselves, never merely as means to another's ends. This mandate receives its sharpest statement in Immanuel Kant's formulation of the categorical imperative in *Foundations of the Metaphysics of Morals,* but the tradition also has roots in both Jewish and Christian traditions. If the embryo is a person, then the deliberate creation of embryos in order to use them "merely as means" to the ends of research appears to violate the ethical mandate not to use persons simply as means to someone else's ends.

These two strong moral convictions—the prohibition on deliberate destruction of innocent persons and the prohibition on use of persons as means to others' ends—raise troubling issues for stem cell research. I share the conviction that it is important to sustain both the ethical principle against killing people and that against using people merely as means to others' ends. Nonetheless, I support embryonic stem cell research. In taking this stand, issues of context are important to my deliberations. Two aspects of the context of the early embryo are of particular importance.

IN VITRO FERTILIZATION: A CRITICAL DIVIDING LINE

IVF has developed into a widespread practice in the United States and elsewhere. Some religious groups (notably the Roman Catholic Church[2]) have raised objections to this practice from the beginning, but the U.S. Congress has neither objected to the practice nor tried to prevent it. Indeed, the practice goes largely unregulated, since it has developed in the private sector. The general citizenry also appears to accept, if passively, the practice of IVF. Three professionals with considerable experience in bioethics have recently argued that "many surplus embryos already exist within in vitro fertilization clinics around the country" and that "research on embryonic stem cells could begin here," using these surplus embryos.[3] In making this argument, these authors appear to accept relatively uncritically the current practice of IVF.

IVF as currently practiced involves both the deliberate destruction of embryos and the deliberate creation of embryos as means to another's ends. Multiple embryos are deliberately created precisely to serve the "ends" of a woman's or couple's desire for a child. Certainly any surplus embryos exist only so the woman or couple (or their physician acting as their agent) may choose an embryo for implantation.[4] Surplus embryos will not become this couple's children. While some may be frozen for possible future implantation, most will be destroyed.[5] If genetically inferior, they will certainly be destroyed. Hence, they are created and used simply to satisfy someone else's ends. We have here a context, then, that involves the deliberate creation of embryos as means to another's ends; acceptance of the practice of in vitro fertilization (IVF) demonstrates that the general public already does accept both the deliberate destruction of embryos and the deliberate creation of human embryos as "means" to another's ends. If embryos are "persons" who should not be killed or used as means only, then IVF should be rejected as a practice.

Indeed, only a few embryos have been destroyed in stem cell research compared to the hundreds that have been created and destroyed in the practice of IVF. Is there something so different about the context that the deliberate creation and destruction of embryos should be permitted in the one case but not in the other?

Changes in the practice of IVF could avoid the problem of deliberate creation and destruction of embryos, whereas stem cell research appears to be impossible without deliberate destruction of embryos; and advances in histocompatibility almost certainly will require some deliberate creation of embryos.[6] In IVF, clinics could fertilize only one or two eggs from each woman and implant every fertilized egg. Such a change in practice would avoid the creation and destruction of surplus embryos. Anderek et al. propose that "we should take as a principle that no human embryo be intentionally 'conceived' (in whatever manner) without some expectation of developing on its own." If they are serious about this principle, then they should argue for changes in the practice of IVF to ensure that no "surplus" embryos are created, for surplus embryos have little expectation of developing. No embryo develops "on its own." Implantation is the sine qua non for the development of embryos. Only those embryos lucky enough to be chosen for implantation have the chance of developing. By accepting the creation of "surplus" embryos, Anderek and his colleagues appear to vitiate their proposed principle that all embryos should have an expectation of developing.

Requiring that all fertilized eggs be implanted would not permit the use of IVF to avoid genetic defects in embryos. It therefore might increase the practice of later abortion of fetuses with genetic anomalies, which, if the fetus is a person, is morally problematic. Some of the purposes currently served by IVF as a practice could not be served if the practice were changed to permit the creation of only those embryos to be implanted.

Alternatively, all fertilized eggs that are not implanted could be frozen. While embryos would still be created and used to serve others' ends, at least they would not be directly destroyed by human agency. If avoidance of destruction of embryos is the goal, then their continued existence, even in an altered state, would appear to be ethically superior to destruction. The practice of freezing embryos, however, raises other ethical questions. Most embryos that are or would be frozen in such a practice will not subsequently be implanted. Indeed, it is commonly agreed among IVF practitioners that, after seven years in a frozen state, they are no longer "living." One could argue that frozen embryos have been consigned to a "limbo" state that is worse than death, theologically speaking, since they have been deprived of baptism and burial. More to the point, however, we must ask: if preservation, even in an altered state, is better than destruction, could one not argue that the derivation of stem cell lines also represents an "altered state" in which surplus embryos that will not be implanted may have ongoing existence, albeit in a new form?[7] Indeed, if the embryos that are used in stem cell research are from "surplus" embryos that will not be implanted and are therefore consigned to death, one might argue that it is more respectful of the value of their life to continue that life in new form by turning those embryos into stem cells rather than simply throwing them in an incinerator.

In short, it is difficult to see why IVF, with its concomitant creation and destruction of embryos, should be accepted, while stem cell research is condemned because it involves the deliberate creation or destruction of embryos. If stem cell research violates moral prohibitions regarding the creation and destruction of persons, it appears that IVF violates those same prohibitions. Like many feminists, I am largely a critic of the practice of IVF.[8] In pointing out parallels between IVF and stem cell research, I am not arguing in favor of IVF. Rather, I argue that it is inconsistent to accept readily—as Anderek et al. appear to do and as the general public appears to do—the practice of IVF and then reject human embryonic stem cell research. Both are practices that involve the creation and destruction of human embryos. Unless IVF is either rejected completely or revised very substantially, it is inconsistent to accept IVF but refuse stem cell research. Acceptance of the creation and destruction of embryos in IVF suggests that the moral question requires a formulation different from that which has thus far permeated the stem cell debate, and I shall make a few comments directed toward a reformulation of the issue.

IS THE CONTEXT ALL-IMPORTANT?

IVF serves the purpose of giving women or couples children, while stem cell research serves the purpose of providing crucial data for basic scientific knowledge of human development. In the long run, such knowledge may be important for developing cures or approaches to devastating conditions such as Alzheimer's disease or paralysis. Nonetheless, the "good results" are further away in time than is the case with IVF; and they are general in character, not focused on the specific desires of a particular couple. IVF serves concrete, immediate needs; stem cell research serves general, future needs. Is this difference morally relevant?

Further, in a culture in which the bearing of children is taken as a basic "right," it may seem that the creation, use, and destruction of embryos is permissible to serve this right, though it would not be permissible to serve other ends. John Robertson has gone so far as to declare that the "right" to have children is so fundamental that individuals have rights of access to any and all technologies that would permit them to have, or to refuse, children, as well as to choose the characteristics of their children.[9] Childbearing appears to hold a somewhat privileged position in U.S. culture, whereas basic research does not. Does this make a difference?

While there may be important differences between the effort to bear a child and the effort to study early human development with a view to locating the keys to "regenerative medicine," it is not clear that these differences should justify setting aside ethical principles in one case but not in the other. President Bush, in his address on stem cells, declared that noble ends do not justify any and all means. He was right. As noble an end as childbearing may be, it does not automatically justify any and all means to this end.[10] If embryos are persons who should be protected against destruction and not used simply for others' ends, then IVF violates important moral precepts, as Roman Catholics and some others have consistently argued. If ends do not justify means, then the importance of the ends—even the fact that they are taken as a basic "right" by many—is not sufficient to justify means that violate fundamental ethical principles.

THE TYPICAL COUNTERARGUMENT

So strong are moral prohibitions on killing or "using" persons, that those who would permit stem cell research typically argue that the principles do not apply because the embryos or blastocysts that would be created or destroyed in research are not "persons" in the moral sense. If embryos are not persons, then the prohibition against killing persons does not apply. If embryos are not persons, then the prohibition against using persons as means to others' ends does not apply. There are good reasons to think that embryos—at least at very early stages such as the blastocyst stage—are not persons. Personhood implies individuation, and the blastocyst has not yet reached the stage where twinning is impossible and individuation guaranteed. Personhood also implies some brain function, and the blastocyst has not developed the neuronal substrate for brain function. Using such criteria, many take a "developmental" view in which the blastocyst is not morally a person.[11]

However, this view presumes that the ethical issues are solved once the status of the blastocyst is resolved. Only two options are considered: (1) the embryo is not a person and therefore stem cell research is permissible; or (2) the embryo is a person, and therefore stem cell research is wrong. The structure of this framing of the argument is precisely the same as the typical structure of arguments around abortion—the fetus is not a "person" and abortion is acceptable, or the fetus is a "person" and abortion is wrong—and it is therefore no wonder that the stem cell debate has become rather hopelessly entangled with the abortion debate. I argue that the structure of these arguments is flawed: determining whether or not an embryo or fetus is a "person" does not settle the moral issue.

REFRAMING THE DEBATE: PERSONHOOD AND MORAL ACTION

A redwood tree is not a person, but this fact does not make it morally acceptable to kill a redwood tree. Conversely, there may be circumstances in which killing persons is morally acceptable—for example, in self-defense or in a "just war." The status of personhood alone does not determine when destruction is permissible.

However, in classic just war theory, the one targeted to be killed is classified as an aggressor; by contrast, "innocent" human life is not to be deliberately taken.[12] Are there any circumstances in which "innocent" people may morally be deliberately killed? A recent case raises the issue with even more poignancy. The case involved conjoined twins where one twin was seriously ill and was threatening the life of both. To separate them meant sure death for the one, but not to separate them meant sure death for both. The only way to save the one was deliberately to take an action that resulted directly in the death of the other. Surely here is an instance where only the most stalwart advocates of "thou shalt not kill" would argue that separating the twins was wrong. Deliberate destruction of one innocent life was necessary in order to save another. Sitting back and doing nothing meant death for both twins. Under these circumstances, most commentators argue that it is acceptable—even morally mandatory—to save the life of the one. Thus, under some circumstances, even the most "innocent" of lives can be directly taken.

The circumstances here involved sure knowledge that the sick twin would die in any case. In one sense, one could say that it was destined for death, and that the only moral issue was whether it was acceptable to hasten its death in the hopes of saving its twin. To some extent, the situation with stem cell research is parallel: the blastocyst not slated for implantation will die anyway, and so the only question is whether it is morally acceptable to hasten its death for the sake of possibly saving other lives.

Many of us recoil at the notion that killing is justified because someone will die anyway.[13] We value human life even when it is slated for death; for example, we do not permit prisoners to be used as research subjects without their will just because they are condemned to die. We have a long tradition of resisting "active euthanasia" even when a patient is dying and is suffering greatly. Why, then, should we permit early embryos to be put to death in research or in IVF just because they are slated to die anyway?

Adult subjects—even those condemned to death—must consent to participate in any research. As self-determining human beings, they cannot be used against their will or without their consent. Early embryos are not self-determining. Moral prohibitions based on self-determination do not apply. It would still be wrong to put that embryo to death if doing so caused pain, since causing pain to sentient beings is wrong.

Concern for the suffering of the sick twin did influence some commentators who argued that the separation was wrong. However, the blastocyst is not a sentient being. When a blastocyst will die anyway, therefore, there is no moral reason that it cannot be used as a research subject, even in research that directly causes its death.

If there are no reservations based on autonomy, and none based on the suffering of a sentient being, then the only reservation that appears to remain is the concern not to be the direct cause of the death of a person. Two things are at stake here.

First, it is sometimes presumed that any direct killing of persons is automatically wrong. But this presumption is precisely what is at issue. The question is what it means to treat people with respect. While we generally think that it is not respectful deliberately to kill a person, there can be circumstances in which it is precisely respectful—for example, to shoot a person trapped in a burning building so that they die a less painful death than they would in burning to death. Hence, we cannot begin with the presumption that direct killing is always wrong, for this is precisely what must be investigated. It is wrong when it violates autonomy, causes pain and suffering, or cuts off the person's destiny. Where there is no autonomy, no sentience, and the person's destiny is death under any circumstances, then direct killing is not wrong.

Second, there are issues of moral integrity at stake. Many people think that if they sit back and "do nothing" or "let nature take its course," then they are not complicit in any wrongdoing. It is better, in this view, to permit a great evil but not to be the cause of a small evil. Thus, some would argue that it is better to watch both conjoined twins die than to step in and save one by a direct action. I disagree. To sit back and watch both twins die when one could be saved is a selfish act. The actor is more concerned with keeping his or her own hands clean than with saving and protecting others. Speaking as a Protestant theologian, I cannot condone this. Jesus gives us the example of one whose love for God caused him to break many rules of "cleanliness" in order to heal and save on the

Sabbath. Morally speaking, what matters most is not whether my hands are clean but whether life is served.

Thus, I argue that the direct killing of the blastocyst in stem cell research is not wrong. The reason it gives pause is that it raises the specter of killing of innocent people. It is good to have reservations wherever such killing is considered. But the prohibition against killing innocent life must be nuanced by looking carefully at the condition of that life. Where there is no autonomy and no sentience, and where death is the only "destiny" of the person, direct killing may be morally countenanced. Failure to do so may reflect moral squeamishness about our own innocence or complicity in evil, but it does not reflect an accurate reading of the ethical principles and the circumstances to which they apply. Destruction of blastocysts for stem cell research may thus be countenanced.

However, this does not solve the problem of creation of embryos for research purposes. While Anderek and his colleagues appear to see no inconsistency between the practice of IVF and their proposed principle that no embryo be conceived without some expectation of developing, I do see a contradiction there. If no embryo should be conceived without some expectation of developing, then neither IVF as currently practiced nor the creation of embryos for stem cell research should be permitted.

But is it true that no human embryo should be conceived without some expectation of developing?[14] Is Anderek's principle sound? Anderek and his colleagues offer no support for this principle. It stands against principles articulated and used elsewhere. Current practice in research on embryos, including stem cell research, operates on a principle that no embryo will be permitted to develop beyond fourteen days. This principle tries to ensure that any embryo that will develop has not been harmed by research. Many people currently believe that no human cloning should be permitted because to attempt to develop a living child by nuclear transfer techniques is to risk serious harm to the child. Rather than adopt the principle proposed by Anderek, we might do well to adopt an opposite principle: that all human embryos that are to develop should be created only by those means that we have reasonable expectation will present no harm to the embryo beyond the "normal" harms of ordinary development in the womb. Such a principle would raise serious questions about the practice of IVF, but would pose no threat to the practice of stem cell research, in which embryos do not develop beyond the early blastocyst stage. In stem cell research, the very fact that the embryos will not develop provides protection against harming any child-to-be, a protection that has not been provided in IVF and would not be provided in reproductive cloning. Thus, on the alternative principle that I propose, it is worse to permit IVF than to permit stem cell research.

CONCLUSIONS

My arguments here will not be satisfactory to those who simply hold that all human life—at any stage or in any context—must be protected against destruction. Very few hold this position, however. Many who condemn abortion nonetheless support "just war" even under modern conditions in which innocent persons will be knowingly and directly killed. My argument for stem cell research is not utilitarian: I did not argue, as a good utilitarian would, that there is more good to be done from stem cell research, which has the potential to help millions, than from IVF, which helps hundreds or at most thousands. My argument is based on principle. Ethical principles require both protection against killing persons and protection against using people simply as a means to one's own ends. These principles must be applied consistently. I argue that it is inconsistent to apply them to stem cell research while ignoring them in the case of IVF, and that there is nothing about the context that would support a different application in the two instances. This leads me to suggest that, if IVF is indeed acceptable as a practice, it points us to the need to reformulate the ethical issues by attending to the arguments that give us these ethical principles. Here, it seems to me that there are instances in which most people would find it ethically acceptable—even mandatory—to kill an innocent person. This suggests that the underlying principle is not one of "no killing," but one of "respect for persons." The question is what respect for persons requires. Where there is no autonomy and no sentience, respect for persons may not require an absolute rule against killing or against using others as means only.[15] Stem cell research does not violate the principle of respect for persons.

NOTES

1. I am indebted here to Dr. Ernle W. D. Young of Stanford University, who has insisted throughout the discussion of stem cells that a crucial dividing line was crossed with IVF. In this essay, I am pursuing some of the implications of that dividing line.

2. The Vatican's Congregation for the Doctrine of the Faith has objected explicitly and strongly to IVF, on grounds both of the destruction of fetuses and of the violation of the intrinsic meaning of the sexual act. See Congregation for the Doctrine of the Faith, "Instruction on Respect for Human Life in Its Origins and on the Dignity of Procreation," in *The Ethics of Reproductive Technology,* ed. Kenneth D. Alpern (New York: Oxford University Press, 1992), 83–95.

3. William Andereck, Paul B. Hoffmann, and David C. Thomasma, "Can Citizens Stem the Biotech Tide?" *San Francisco Chronicle,* 5 August 2001, D8. I would point out that research on stem cells did indeed "begin here," with the so-called "surplus" embryos from IVF.

4. The more correct term here is probably "introduction," since it is never guaranteed that every embryo introduced into the womb will in fact implant. However, I follow popular parlance here and use the term "implantation."

5. According to Carl T. Hall, "The Forgotten Embryos," *San Francisco Chronicle,* 20 August 2001, A11, more than 50 percent are immediately "discarded."

6. I have inquired as to whether stem cells might be derived by removing a few cells from the inner cell mass rather than destroying the trophoblast and culturing the entire inner cell mass; and the scientific consensus appears to be that it is not possible, as it takes the cluster of cells in order to get cell development and growth in culture.

7. This appears to be akin to an argument made by Glenn McGee, who suggests that if the genetic status of the embryo is its defining charac-

teristic, rather than its "hollow ball" physiology, then it has not been destroyed in being turned into stem cells, because every stem cell is a living continuation of that genetic status.

8. For example, clinics often measured their "success" rates by how many times they were able to implant an embryo, regardless of whether the embryo came to term. From the perspective of women seeking not simply the experience of pregnancy but also the birth of a live child, announced "success" rates were therefore deceptive, since the birth of live children was considerably less than the rate of implantation. I would argue further that IVF as a practice reinforces stereotypical views of women's roles, making the capacity to bear children too central to women's identity.

9. John A. Robertson, *Children of Choice: Freedom and the New Reproductive Technologies* (Princeton, N.J.: Princeton University Press, 1994).

10. Indeed, if one were a good utilitarian, one would argue that it seems very strange to permit moral rules to be set aside for the sake of the desires and interests of one or two people, but not to permit them to be set aside for the sake of the health and interests of millions.

11. I believe this view has a strong moral foundation and, therefore, I joined colleagues in accepting this view as the basis for a report on the permissibility of stem cell research. See Geron Ethics Advisory Board, "Research with Human Embryonic Stem Cells: Ethical Considerations," *Hastings Center Report* 29, no. 2 (1999): 31–36.

12. To be sure, the classic distinction between killing aggressors and innocents becomes problematic under modern conditions using weapons of mass destruction, since these weapons will kill innocent civilians as well as military targets. The ratio of civilians to military personnel killed in the Second World War is estimated to be twenty to one—an indication that in modern warfare one cannot claim that only "noninnocent" persons will be killed.

13. However, Laurie Zoloth notes that Jewish halachic reasoning permits different treatment once it is clear that a person has a fatal organic illness. See Zoloth, "The Ethics of the Eighth Day," in *The Human Embryonic Stem Cell Debate*, ed. Suzanne Holland, Karen Lebacqz, and Laurie Zoloth (Cambridge, Mass.: MIT Press, 2001), 99.

14. Strictly speaking, Anderek and his colleagues frame the principle as one of "developing on its own." This is nonsense, as no embryo develops on its own. For development, the proper environment is necessary. Women's wombs are still the sine qua non of embryonic development.

15. I have argued at more length about the requirements of respect for persons in "The Elusive Nature of Respect," in *The Human Embryonic Stem Cell Debate*, ed. Holland et al.

CHAPTER 7

Levels of Moral Complicity in the Act of Human Embryo Destruction

EDWARD J. FURTON

One of the most pressing and difficult moral questions that will face Catholic health care in the future is whether physicians and patients may make use of any new and promising therapies that might derive from research on embryonic stem cells. Although the prospect of curative therapies for Alzheimer's, Parkinson's, and other serious diseases is nothing more than a promise at present, if these researches are successful, Catholic health care facilities will have to determine whether their use would implicate them in the immorality of human embryo destruction. If it does, then it will be necessary for Catholics and Catholic health care facilities to forgo these therapies regardless of whatever benefits they may provide to suffering patients. Obviously, this would put Catholic health care in a dramatically negative light—one in which many would accuse it of abandoning the best interests of patients.

Fortunately this potential nightmare is only a scenario, not a certainty. There is the very real promise of adult stem cell research, which is not only above reproach from a moral perspective, but which also has several scientific advantages over embryonic stem cell research. However, in my view, Catholic health care facilities, Catholic physicians, and Catholic patients alike should be able to make use of therapies that derive from embryonic stem cells despite their immoral origins. My judgment in this matter rests upon certain facts about the nature of cell lines and their capacity to grow in culture and also upon a moral analysis that relies upon the principles of cooperation.

The principles of cooperation remain a controversial area that is routinely subject to serious disputes among philosophers and theologians. In spite of this controversy, it is clear that these principles are the appropriate tools for analysis in this case. Each question discussed below revolves around whether the use of embryonic stem cells would constitute an immoral association with the original act of embryo destruction. Does present use of this new therapy make me complicit in that wrong? I take it as a given that all Catholics would agree that human embryo destruction is immoral, although some Catholic philosophers and theologians deny this. My aim here is not to enter the debate over "immediate or delayed hominization," but to accept it as a given that, even if one holds that the embryo is not a human being (highly implausible, in my view), one is nonetheless obliged to take the safer course and refuse to participate in any act of embryo destruction.[1]

THE PRINCIPLES OF COOPERATION

My view of cooperation will follow that of McHugh and Callan, as expressed in their classic work *Moral Theology.*[2] Suffice it to say that I do not accept the "proportionalist" school of moral theology, which tends to deny the existence of intrinsically immoral acts (or denies the existence of exceptionless moral norms—sometimes calling them "virtually exceptionless"). In this school of thought, certain evils are said to be "pre-moral evils," meaning that they are evils prior to any particular intention or circumstances. They may or may not become moral evils depending upon the intention of the agent and the circumstances he faces. Thus human embryo destruction may be described as a "pre-moral evil" that becomes an actual moral evil when joined to the intention of this particular agent or these particular circumstances. I do not believe it is helpful to enter into the particulars of this debate, in large part because I am interested in providing an analysis that will be helpful to those in the health care profession. Debates over theological method distract from that aim.[3] The other difficulty is that, with the publication of Pope John Paul II's encyclical letter *Veritatis Splendor,* the mere mention of the word "proportionalist" has become highly charged with questions of religious orthodoxy.[4]

In my view, human embryo destruction is moral evil by its very nature (or because of its object) and regardless of the intention of the agent or the circumstances. By "intrinsically evil" I mean an action that has "no good use," as McHugh and Callan say, or that is immoral in each and every case; by "object" I mean the intelligible structure of the act as it arises from within the social order and the natural relations of justice that ought to comprise it. This description is fully sufficient for the purposes at hand. The question that faces us here is whether one who receives a benefit from the act of human embryo destruction is in any way responsible for that wrong or cooperates in an intrinsically evil act.

Let me set forth some of the standard distinctions of cooperation, as presented by McHugh and Callan. Those familiar with this subject, or who are not interested in such technical matters, may wish to proceed immediately to the next section. For the remainder, we may begin with a general definition of cooperation. "Cooperation or participation in sin, strictly understood, is help afforded another, whom one has not seduced, to carry out his purposes of sinning."[5] Formal cooperation arises when one shares in the evildoer's intention. Formal cooperation is explicit "when the end intended by the cooperator (finis operantis) is the sin of the principal agent." Thus one who intends the same immoral act as the principal agent—for example, the destruction of a human embryo—engages in explicit formal cooperation.

Formal cooperation can also be implicit, "when the cooperator does not directly intend to associate himself with the sin of the principal agent, but the end of the external act (finis operis), which for the sake of some advantage or interest the cooperator does intend, includes from its nature [i.e., its object] or circumstances the guilt of the sin of the principal agent."[6] We need to parse this complex sentence. Here cooperation does not result from a sharing in the same intention, but from a willingness to participate in the same activity, though the cooperator will claim to do so ostensibly for some other purpose than that intended by the principal agent. Thus an individual might be strongly opposed to the destruction of human embryos, but nonetheless provide direct assistance to that activity because he believes it will lead to new cures for serious diseases. He sees his cooperation as a means to some further end and is willing to overlook or "tolerate" an evil in view of that good. There is no objective difference between explicit and implicit formal cooperation, because to will the good end (the potential cure of some disease) it is necessary to will the immoral means (the destruction of human embryos).

Material cooperation occurs when one "does not intend the sin whose external commission one is aiding," and it may be either proximate or remote depending upon the nearness and the definiteness of the assistance provided.[7] Thus proximate material cooperation by means of nearness would occur if a researcher held the petri dish into which the embryonic stem cells from the destroyed human embryo were placed; but there would be remote material cooperation if the cooperator simply handed the dish so that the principal agent could place the cells there himself. Proximate material cooperation by means of definiteness would occur if one installed the needle that would be used in the laboratory machinery to pierce the trophoblast of the embryo, but there would be only remote cooperation if one was supplying needles to the laboratory as a vendor of laboratory supplies.

Material cooperation may be blameless if there is a sufficiently weighty reason for carrying it out.[8] Without such a reason, however, the action is immoral. McHugh and Callan hold that the "first condition of [legitimate] material cooperation is that the act of the cooperator must be good or at least indifferent; for if it is evil, the cooperation becomes implicitly formal."[9] We judge whether an action is evil by examining its nature (i.e., its object) and its circumstances. An action is not only evil, but evil by its very nature (intrinsically evil) "if it has no uses except such as are evil; it is indifferent, if, according to the intention of those who use it, it is now good, now evil."[10] The destroying of human embryos is intrinsically evil, but in contrast, the handing of a petri dish to another researcher is an indifferent act, for the dish can be used for either good or bad purposes.

The activity of handing over a petri dish may not implicate one in wrongdoing on the grounds of nearness, but we must also weigh this contribution in terms of its circumstances; for material cooperation is "evil, if by reason of adjuncts it is wrong, as when it signifies approval of evil, gives scandal to others, endangers the faith or virtue of the cooperator, or violates a law of the Church."[11] One who works alongside a researcher who destroys human embryos, even though he does not directly participate in that act of destruction itself, does all (or at least most) of these things. He appears to give his approval to that destruction; causes others to doubt his moral convictions; and, if he is a Catholic, he endangers the faith of those who cannot understand his association with this project. He may also violate the church's prohibition on any association with the intentional killing of innocent human beings.[12]

There is one last distinction among types of cooperation that must be considered here: immediate versus mediate material cooperation. This distinction has proven to be the most controverted of all.[13] According to McHugh and Callan, material cooperation is either immediate or mediate "according as one shares in the sinful act of the principal agent, or in some act that preceded or followed it."[14] Thus the researcher who looks through a microscope as his partner destroys the embryo, in order to ensure that the needle is properly positioned, engages in immediate material cooperation. His action is contemporaneous to that of the principal agent and essentially joined to it. If he provides no such assistance, but prepares the machinery prior to its use or takes care of it afterwards, then his cooperation will be mediate material cooperation.

THE CENTRAL QUESTIONS

In view of the above presentation of the principles of cooperation, I want to raise the following three questions:

1. May a Catholic researcher or scientist receive embryonic stem cells after the human embyro has been destroyed, in order to establish an embryonic stem cell line?
2. May a Catholic researcher or scientist use an established embryonic stem cell line in order to discover new drugs or therapies to cure debilitating diseases?
3. May a Catholic physician or patient make use of a new drug or therapy that has been derived using an embryonic stem cell line for the cure or treatment of a disease?

My responses to these three questions, in general, will be as follows. (1) No, a Catholic researcher or scientist may not work to establish an embryonic stem cell line, because his work would constitute immediate material cooperation in the destruction of human embryos; (2) No, a Catholic researcher or scientist may not work with an established embryonic stem cell line in order to discover new drugs or therapies; to do so would constitute proximate material cooperation with the intrinsically evil action of human embryo destruction; (3) Yes, a

Catholic physician or patient may make use of any new drug or therapy derived from an embryonic stem cell line, because the level of cooperation would be remote at best.

Before proceeding to defend the argumentation in support of these conclusions, let me say that the whole of my analysis depends upon seeing cooperative action as a causal contributor to evil, whether that evil is the act of embryo destruction itself or some evil that follows upon that contribution: for example, encouraging others to destroy human embryos. To cooperate is to be causally responsible (at least in part) for the evil action of the principal agent. If the cooperator adds no causal dimension to the wrongful action of the principal agent, then either the cooperator's contribution is above moral reproach or it is not cooperation at all. Either way, the cooperator is not responsible for the wrong that is done. If there is a causal contribution to the wrong committed by the principal agent, then it can only be justified if the level of cooperation is proportionate to the importance of the good to be gained or the gravity of the evil to be incurred.

THE VACCINE PARALLEL

My reasoning in this matter follows an earlier opinion given concerning the case of vaccines that have a distant origin in acts of abortion.[15] It is interesting to note that President George W. Bush referred to this view, in a general way, when justifying his decision to allow federally funded research on embryonic stem cells.[16] There are a number of vaccines currently on the market that are grown in the human cell lines MRC-5 (Medical Research Council 5) and WI-38 (Wistar Institute 38). These lines had their origin in lung tissue taken from two separate acts of abortion occurring several decades ago.[17] The question facing conscientious individuals who do not want to cooperate in any way with abortion is whether, if they use these products, they are implicated in that original act of destruction. This poses a particularly serious problem for parents who learn of the association between a certain vaccine and these abortions at the time that their child is about to be immunized.

In many cases there are vaccines of equal medical efficacy on the market that do not have these associations; but in several cases, there are no alternatives.[18] This means that one either uses a vaccine with a questionable origin or avoids immunization altogether. There is no middle path. The right to refuse vaccination on either religious or philosophical grounds is recognized in almost all states; but, in my view, parents should immunize their children even if there are no other products available other than these problematic vaccines. Why take this view? There are a number of reasons.

First of all, these abortions did not occur for the sake of making the cell lines. They were carried out "for therapeutic reasons" (a euphemism for elective abortion prior to *Roe v. Wade*); and then, after that destruction was completed, lung tissue was retrieved in order to start a cell line. So the problem of sharing in the original evil intention is eliminated. Second, the descendant cells from the extracted tissue are not the same as those that were removed from the abortion. The vaccine that the patient uses is not grown in the lung cells of the aborted child, but in descendants that have grown in culture following the abortion. Once a cell has divided into two daughter cells, neither can be called the same as the original cell. In fact, the original cells from the abortion no longer exist in the cell lines. Within a short period of time they are completely replaced with daughter cells. This is a point that is sometimes misunderstood by parents who believe that the vaccine contains the actual cells of the aborted child.[19] Third, the cell line is self-propagating. That means that use of the cells to make the vaccines will not encourage the future practice of abortion because the cells replenish themselves and are available in almost limitless supply.

We want to avoid wrongdoing, of course; but, in addition to abiding by negative moral norms, we also have positive moral duties. These must be fulfilled whenever possible. One such duty is the responsibility to promote our own health and that of our neighbors—part of our duty to the common good. In the case of our children, this duty becomes paramount because parents must make medical decisions on behalf of their children who are not yet at the age of reason. The complications that arise from diseases like measles, mumps, and rubella can be very serious, resulting in permanent injury or death. We also know that the more individuals refuse vaccinations, the greater the likelihood that these diseases will spread and affect the whole of society, including those who have already been vaccinated.[20] Parents must weigh any concern about a possible association with abortion against the risks to the health and life of their children. This is not a case of "choosing the lesser evil," but of choosing a good that has only, at best, a distant connection with a past evil.[21]

If an adult wishes to make a strong statement against the practice of abortion, then I think he should feel free to refuse immunization with these vaccines, but there does not appear to be any moral obligation that he do so. Refusal would be an act that goes beyond the moral norm. Although it does place others at some risk, and thus potentially harms the common good, it seems reasonable to say that the harm is sufficiently outweighed by the evil that one is opposing. The act is a heroic one for a good cause. The situation is quite different, however, when it comes to a child. Here the moral obligation to preserve the life and health of a child should take precedence over any concern about a distant association with abortion. The parent does wrong, in my view, if he will not allow the child to be vaccinated; for he places the child's health and life at risk when there is no overriding moral obligation to do so.

APPLICATION TO EMBRYONIC STEM CELL RESEARCH

Let us now return to the three questions raised above and apply this line of reasoning to them. The one who destroys human embryos, obviously, is directly responsible for that act of wrongdoing. He kills an innocent human being and thereby harms an inviolable good—a good that should never be directly attacked for any reason. In the terms of cooperation enunciated by McHugh and Callan, he is the "principal agent." Anyone who assists the researcher by sharing in the intention of this act of destruction would be engaged in "formal cooperation." Nor does it matter if he claims to do so for the sake of curing serious diseases. As McHugh and Callan say, this is implicit formal cooperation.[22]

One difference between the vaccine case and embryonic stem cell research is already apparent. In the case of the vaccines, the abortions were not performed for the sake of obtaining cellular material to start new cell lines (that happened, as it were, after the fact); but in embryonic stem cell research, the embryos are destroyed solely for the purpose of supplying their cells to researchers. This adds a higher degree of moral gravity to the case and makes dissociation from the original act of wrongdoing much more difficult. A stamp or imprint is left on the whole field of research because of the motives that originally gave rise to the killing. Until there is a complete ban on all human embryo destruction in the United States, any scientist or researcher who associates with embryonic stem cell research will do his work under a moral cloud.

ESTABLISHING CELL LINES FROM DESTROYED EMBRYOS

Once the embryo has been destroyed, the inner cells are placed into glass and encouraged to grow into an immortalized cell line. Whether receiving lung tissue from an aborted fetus or the inner cells from a destroyed human embryo, the researcher is working with the living remains of a destroyed human being. These tissues and cells have a life of their own that can be distinguished from the life of the fetus or the embryo; nonetheless, they should be treated with the respect that all human remains deserve.[23] May a Catholic researcher or scientist receive these inner cells from the destroyed human embryo and attempt to start an embryonic stem cell line with them? I presume here that the one who receives the inner cells from the destroyed human embryo is not the same person who carried out the destruction—though this is not only possible, but very likely to be the case. Most of those who work to establish embryonic stem cell lines are the very same people who destroy the embryos.[24] What I want to consider, however, is the case of the moral agent who recognizes that human embryo destruction is immoral, but who also wishes to participate in promising scientific research on these cells for the sake of finding a cure for serious diseases. Such a person does not share in the intention of the one who destroyed the embryos, but only begins to participate in research after the fact of embryo destruction. Hence, it is clear that his actions do not constitute formal cooperation.

Despite his good intentions, it seems to me that his actions are subsumable under the heading of immediate material cooperation. As McHugh and Callan indicate, it is never possible to cooperate with an intrinsically evil action. Working with the remains of a human being who has been unjustly killed by another is participating in the wrongful act of killing, if there is any coordination of their activity. In this case, the two must coordinate their respective actions very closely. The Catholic researcher or scientist would have to agree to receive the remains at a certain time and place, work to ensure that the materials do not spoil, prepare his own lab for receipt and preservation of the cells, and the like. This shows foreknowledge of the evil to be done and a practical willingness to accommodate it.

The case is similar, it seems to me, to that of the researcher who conducts experiments using fetal tissue from abortions, such as those who inject fetal cells into the brains of patients in an attempt to cure Parkinson's disease.[25] In both cases, one is working with the actual remains of another human being and coordinating one's activity in order to ensure that the tissues or cells are received in the best possible condition for research purposes. Coordinating one's actions with those who extracted the cells from the destroyed human being implicates one in an intrinsically evil act. So long as any of the original cells from the destroyed human embryo remain in the petri dish, it seems to me that the one who handles them engages in immediate material cooperation.[26]

WORKING WITH ESTABLISHED EMBRYONIC CELL LINES

The situation begins to change once a cell line has been established and the embryonic stem cells are growing in culture. For, at that stage, the cells replicate themselves independently of the original cells taken from the destroyed human embryo. Although the inner cell mass that was used to begin this cell line was once a part of an existing human being, the "daughter" cells that are made available for general scientific use are not. They appear in culture only after the death of the embryo proper and the replacement of the original cells of the embryo with daughter cells. Though they contain the same DNA, these daughter cells are not governed by any formal principle that unifies them into any larger organized whole, as was the case with the destroyed human embryo. Each descendant cell has its own identity and together they form a disorganized mass.

This leads to the second question: Is it possible for a Catholic researcher or scientist to use one of these established embryonic stem cell lines in the search for new drugs and therapies to cure debilitating diseases? President Bush's decision made several such lines available for use in federally funded research programs. In my view, once a cell line has been established and there are no longer any of the original cells of the embryo remaining in glass, the Catholic researcher is not engaged in action that is intrinsically evil. Nonetheless, these are not just any cell lines. The conscientious scientist who works with these cells must face the fact that his research could not proceed unless someone had destroyed human embryos. As soon as he begins his work, he becomes a part of a larger research community that includes others who are carrying out the morally repugnant action of destroying human embryos. Any progress that the Catholic researcher makes will lead others to conclude that it is good to destroy human embryos. Thus, if he finds a cure for Alzheimer's disease, others will say that he was right not to be concerned about the evil of human embryo destruction. As we saw above, McHugh and Callan state that cooperation with evil is not permissible "if by reason of adjuncts it is wrong, as when it signifies approval of evil, gives scandal to others, endangers the faith or virtue of the cooperator, or violates a law of the Church." That would appear to be the case here.

In addition to this problem, there is another helpful parallel to the vaccine case. The problematic cell lines that are currently in use in the production of certain vaccines could have been started using other cells besides human cells. Other vaccines on the market are grown in chick embryos, monkey kidney cells, or some other nonhuman source.[27] It was not necessary to grow these vaccines in human cells at all, though now that this procedure has been established, changing to another type of cell line would be very difficult. The properties of the existing cell lines MRC-5 and WI-38 are well known; scientists are familiar with their use. A change in vaccine production would require additional time and research, as well as review and new approval by the Food and Drug Administration. What is important to note here is that, once the die is cast in favor of using a particular type of cell line, it is very hard to change course. If there is success in treating a disease such as Alzheimer's or Parkinson's using embryonic stem cell research, it will be next to impossible to switch to adult stem cells even if they are clinically proven to hold equal promise.

We presently have an opportunity to begin this field of research aright by setting aside the embryonic stem cell lines and working instead with adult stem cells. Every year there is new evidence that adult stem cells are as promising as the embryonic cells, and, in some ways, even more so.[28] The problem of immune rejection, which has plagued efforts to transplant human organs, will also affect any new therapy based in embryonic stem cells. Adult stem cells do not suffer this problem. They are taken from the body of the patient (e.g., from the bone marrow), reformulated to the correct cell type (e.g., muscle cells), and then injected where they will combine with diseased tissue (such as the weakened walls of a failing heart). There is no possibility of immune rejection because the cells are the patient's own and are simply being moved from one place to another in the same person's body.

Proponents of embryonic stem cell research say that the solution to the problem of immune rejection is "therapeutic cloning," but this is substituting something worse for what was already bad enough. The idea here is to take a body cell from a patient, remove its nucleus, place that genetic material into an enucleated ovum, and produce a clone, who would then be destroyed in order to retrieve embryonic stem cells. These cells would be immune system compatible because they would come from a genetic twin of the patient. This may be a scientific solution, but it is not a moral one. Others say that we can overcome this problem by establishing a large bank of embryonic stem cell lines that will be histocompatible with all members of society, but here the better solution would be to use stem cells from umbilical cords and placentas. If widely collected in hospitals after childbirth, these could fulfill the same purpose.

Although research on established embryonic stem cell lines does not involve direct participation in an intrinsically evil act, I believe that Catholic scientists have an obligation to forgo this type of research. Such research would qualify as mediate material cooperation, but it still remains closely tied to the original act of embryo destruction. McHugh and Callan permit mediate material cooperation when the evil suffered to oneself is grave or at least proportional to that suffered by the injured party.[29] The loss of livelihood or prestige by a refusal to participate in embryonic stem cell research may indeed be significant, but it is hard to see how it could be more significant than the loss of life suffered by the embryo. As with the case of vaccines, whenever there is an alternative to using a product that originates in the destruction of human beings, that alternative should be chosen. Adult stem cell research is a promising alternative to embryonic stem cell research, and Catholic scientists should pursue this field instead. They may find out that they have not only chosen the better path from a moral perspective, but that this field holds far greater medical and scientific promise.

THE USE OF PRODUCTS FROM EMBRYONIC STEM CELL LINES

We take up, finally, the third question: May a Catholic physician or patient make use of a new drug or therapy that has been derived using an embryonic stem cell line for the cure or treatment of a serious disease? Here we consider the possibility that new therapies derived from embryonic stem cell lines will make their way into physicians' offices and into health care centers, including those that are Catholic. We leave the laboratory of the scientific researcher and stand at the bedside of the patient. What are physicians, hospitals, and patients to do in light of these new therapies? Shall they refuse cures for themselves and their patients because of their origin in destroyed human embryos? Presuming that embryonic stem cell research is successful, would it be wrong for those who are Catholic to benefit from them?

Here I think we finally achieve a level of cooperative distance that would enable a person concerned about any inappropriate association with the destruction of human embryos to make use of the fruits of this type of research. Let us suppose that there is an embryonic stem cell line that has been carefully tested and approved for use in medical therapies by the Food and Drug Administration. As a self-propagating line, the use of its cells in a new drug or therapy would not require any further destruction of human embryos. The therapies derived from this line, therefore, would be distinguishable from the larger project of human embryo destruction that is currently underway. Given their isolation from this wider practice, it seems to me that one could make use of these therapies without thereby encouraging others to destroy additional human beings. My use consumes cells only from this line, which is self-replenishing.

As with the case of vaccines, if there were any alternative product or therapy available that would be essentially equivalent to that derived from embryonic stem cells, then the patient should make use of that instead. When there is no alternative to using a product or therapy that has its origin in an embryonic stem cell line, then that use would be morally appropriate if it were for a sufficiently serious purpose, such as curing an illness or saving a life. Such use would constitute remote material cooperation. The individual who uses the drug or therapy obviously does not intend that any human embryos be destroyed; the aim is the cure of a serious illness. The cells that are employed in the drug or therapy are not the original cells taken from the destroyed human embryo, but the descendant cells. Given their presence in an established stem cell line, which is self-propagating and thus generating a virtually unlimited supply of replacement cells, use of the drug or therapy by the patient would not be an incentive to the practice of embryo destruction. Thus the conclusion here is the same as that given above in the case of vaccines.

As with the vaccines, an individual who wished to make a particularly strong statement against the destruction of human embryos might refuse even a lifesaving cure that derived from these lines. Such an action would go beyond what is morally necessary and would be heroic. But again, I believe it would be immoral for parents to deprive their children of curative therapies derived from established embryonic stem cell lines, for there is nothing inherently immoral in their use. The level of cooperation is far too distant to be morally problematic. Their use would secure the goods of life and health for the child and therefore, if such therapies did become available in the future, parents would have a moral obligation to use them for the good of their children. The reason in favor of their use would far outweigh any remote association with embryo destruction.

NOTES

1. In the question of hominization I am happy to take the work of Mark F. Johnson as my guide. His most recent analysis is "The Moral Status of Embryonic Human Life," in *What Is Man, O Lord? The Human Person in a Biotech Age,* ed. Edward J. Furton (Boston: The National Catholic Bioethics Center, 2001), 181–98. See also his earlier and widely cited "Delayed Hominization: Reflections on Some Recent Catholic Claims for Delayed Hominization," *Theological Studies* 56 (1995): 743–63. For an example of a Catholic scholar who adopted delayed hominization but considers all human embryo destruction immoral, see Norman Ford, S.D.B., "The Human Embryo as Person in Catholic Teaching," *The National Catholic Bioethics Quarterly* 1, no.2 (Summer 2001): 155–60.

2. John A. McHugh, O.P., and Charles J. Callan, O.P., *Moral Theology: A Complete Course Based on St. Thomas Aquinas and the Best Modern Authorities,* 2 vols. (New York: Joseph F. Wagner, Inc., 1929); revised by Edward P. Farrell, O.P. (1958). See especially volume 1, nos. 1506–26. See also Thomas Aquinas, *Summa Theologiae,* I–II, qq. 18 and 19.

Some will wonder why I turn to a book that predates so much of the contemporary discussion of cooperation, but this is its chief merit. Moral theology has been a field brimming with controversies over the past forty years. This work, revised before the appearance of a theology that has called into question traditional moral categories (even as it asserts that it remains true to them), suffers none of this confusion. McHugh and Callan are excellent expositors of what were already some very difficult and subtle distinctions.

The principles of cooperation were given their first form under the direction of St. Alphonsus Liguori (1696–1787). An excellent article explaining his ideas and their influence upon Catholic moral reasoning is Roger Roy, C.SS.R., "La Cooperation selon Saint Alphonse Liguori," *Studia Moralia* 6 (1968): 378–435.

3. In brief, I reject any bifurcation of the moral object into "pre-moral" and "moral" dimensions. I consider this bifurcation to be the product of the

"is-ought" distinction of David Hume, which was later incorporated into Kantianism, and which drives so much of modern debate in moral theology.

4. Thus Richard McCormick, commonly described as a proportionalist prior to this encyclical, argued that *Veritatis Splendor* had set up a straw man and rebutted what was not properly proportionalism at all. *Veritatis Splendor* and "Moral Theology," *America* 37 (March 1976), 8–11.

5. McHugh and Callan, *Moral Theology,* no. 1506.

6. Ibid., no. 1511. The end intended by the agent (finis operantis) is sufficient of itself to incur guilt, even if the act itself never occurs; hence the end of the act (finis operis) does not add any essential element to its wrongfulness—even though there is an enormous difference in the real order between an action contemplated and an action completed.

7. Ibid., no. 1512.

8. McHugh and Callan offer the following standards for judging whether a reason is sufficiently weighty (ibid., no. 1519): "the graver the sin that will be committed, the graver the reason required for cooperation"; "the nearer the cooperation is to the act of the sin, the greater the reason for cooperation"; "the greater the dependence of the evil act on one's cooperation, the greater the reason required for cooperation"; "the more evil the act, the greater the reason for cooperation"; "the more obligation one is under to avoid the act of cooperation or to prevent the act of sin, the greater the reason required for cooperation."

9. Ibid., no. 1517.

10. Ibid.

11. Ibid.

12. Thus directive no. 45 of the National Conference of Catholic Bishops' *Ethical and Religious Directives for Catholic Health Care Services* (Washington, D.C.: United States Catholic Conference, 1995) states that "Catholic health care institutions are not to provide abortion services, even based on the principle of cooperation." The same restriction on health care personnel in a Catholic health care facility might apply to any association with human embryo destruction on the part of a Catholic researcher.

13. The controversy had its focus primarily in the ill-fated Appendix to the National Conference of Catholic Bishops' *Ethical and Religious Directive for Catholic Health Care Services* (ibid.). This version (no longer in print and replaced by the revised 2001 edition) held that "immediate material cooperation is wrong, except in some instances of duress." This led a great many Catholic hospitals to claim that they were engaged in legitimate material cooperation when they permitted physicians to carry out procedures that were deemed intrinsically immoral by the church. They claimed "duress" under the (very real) possibility of financial loss or even closure. The problem with the language of the 1995 Appendix, in my opinion, was not that it justified immediate material cooperation under duress, but that

it failed to note the immorality of any type of cooperation with intrinsically evil actions. Thus the Appendix could have avoided a great deal of controversy and accurately reflected the tradition if it had said: "immediately material cooperation with intrinsically evil actions is always wrong—period."

The failure to draw this distinction is rife throughout much of the literature in moral theology. A representative example is James F. Keenan's extremely influential "Prophylactics, Toleration, and Cooperation: Contemporary Problems and Traditional Principles," *International Philosophical Quarterly* 39 (June 1989): 205–20. At page 216 he states: "Generally speaking immediate cooperation is always wrong, though the manualists saw, even here, exceptions. These exceptions are raised in instances of constraint." True enough, but no manualist allows for immediate cooperation in intrinsically evil actions, even when there is constraint (or "duress"). Thus McHugh and Callan allow for immediate material cooperation in a variety of situations (see *Moral Theology,* nos. 1521–23), but also say that "if one cannot cooperate immediately without performing an act that is intrinsically evil, immediate cooperation is, of course, unlawful" (no. 1526).

14. McHugh and Callan, *Moral Theology,* no. 1509.

15. "Vaccines Originating in Abortion," *Ethics & Medics* 24, no. 3 (March 1999): 3–4. Others have taken a similar position. I am indebted to the work of Daniel P. Maher, whose unpublished and undated paper "On the Use of Certain Vaccines," available from the National Catholic Bioethics Center, Boston, has served as a starting point for my own reflections. Maher has recently published a more extensive statement of his views: "Vaccines, Abortion, and Moral Coherence," *The National Catholic Bioethics Quarterly* 2, no. 1 (Spring 2002): 51–67.

16. "There is a precedent. The only licensed live chickenpox vaccine used in the United States was developed, in part, from cells derived from research involving human embryos. Researchers first grew the virus in embryonic lung cells, which were later cloned and grown in two previously existing cell lines. Many ethical and religious leaders agree that even if the history of this vaccine raises ethical questions, its current use does not." George W. Bush, "Stem Cell Science and the Preservation of Life," *New York Times,* 12 August 2001, A13. Mr. Bush's description of how these vaccines are made is not entirely accurate and, as one would expect in a newspaper opinion piece, the nuances of the moral question are glossed over.

17. J.P. Jacobs, C.M. Jones, and J.P. Baille, "Characteristics of a Human Diploid Cell Designated MRC-5," *Nature* 227 (July 11, 1970): 168–70; L. Hayflick and P. Moorhead, "Serial Cultivation of Human Diploid Cell Strains," *Experimental Cell Research* 25 (1961): 585–621; and L. Hayflick, "The Limited In Vitro Lifetime of Human Diploid Cell Strains," *Experimental Cell Research* 37 (1965): 614–36.

18. Vaccines for which there are presently no alternatives are "Varivax" (Merck & Co., Inc.) for varicella, also known as chicken pox (uses both WI-38 and MRC-5 human cell lines); "Havrix" (SmithKline Beecham) for hepatitis A (uses MRC-5); "Vatqua" (Merck & Co., Inc.) for hepatitis A (also uses MRC-5); "Biavax II" (Merck & Co., Inc.) for rubella, also known as German measles (uses WI-38); and "Meruvax II" (Merck & Co., Inc.) for rubella (uses WI-38 line).

19. Vaccines are grown in descendant cells and then removed and purified for use; hence, not even any of the descendant cells remain in the vaccine that is used for immunization. This is quite different from what happens in fetal tissue research. See Maria Michjeda, "Fetal Tissue Transplantation: Miscarriages and Tissue Banks," in *The Fetal Tissue Issue: Medical and Ethical Aspects,* ed. Peter J. Cataldo and Albert S. Moraczewski, O.P. (Braintree, Mass.: The Pope John XXIII Medical-Ethics and Education Center, 1994), 1–14.

20. See, for example, "Individual and Community Risks of Measles and Pertussis Associated with Personal Exemptions to Immunization," *Journal of the American Medical Association* 284, no. 24 (December 27, 2000): 3145–50.

21. Note that it is not the amount of time that has passed that makes this connection "distant," but the reasons given in the third paragraph of this section.

22. Nor does it matter that the parents of the embryo have agreed to allow their offspring to be destroyed. No one has the authority to allow another person to kill an innocent human being.

23. Catholic teaching distinguishes between the death of the person from the life that remains in the organs and tissues for a period of time after death. "Considerations of a general nature allow us to believe that human life continues for as long as its vital functions—as distinguished from the simple life of the organs—manifest themselves spontaneously or even with the help of artificial processes." "An Address of Pope Pius XII to an International Congress of Anesthesiologists," in *Conserving Human Life* (Braintree Mass.: The Pope John Center, 1989), 312–18. The life of embryonic stem cells is similar to the life found in the body of one who has died, but whose organs are still capable of being transplanted. Thus the relatives of a brain-dead person may give their consent to have the organs of the deceased removed for transplantation to others. John Paul II, "Address to the International Congress on Transplants" (29 August 2000), in *The National Catholic Bioethics Quarterly* 1, no. 1 (Spring 2001), n. 5. These remain "alive" with a life that is different from that of the personal life of the deceased. So, too, the inner cells of a destroyed human embryo retain their own life, even though that life is not the same as the personal life of the individual who was killed in the act of embryo destruction.

24. This is the reason why the policy of the Clinton administration, which offered to fund embryonic stem cell research so long as the embryos were not destroyed with federal dollars, was viewed by many as morally bankrupt. A sufficient moral distinction had not been established between the original act of embryo destruction and the later use of the cells.

25. See my "The Nebraska Fetal Tissue Case," *Ethics & Medics* 26, no.1 (January 2001): 3–4.

26. My reasoning here differs slightly from that of the Pontifical Academy of Life and its "Declaration on the Production and the Scientific and Therapeutic Use of Human Embryonic Stem Cells," *L'Osservatore Romano* [English edition], (13 September 2000): 7. Concerning the question "whether it is morally licit to use ES cells, and the differentiated cells obtained from them, which are supplied by other researchers or commercially available," the Academy replied: "The answer is negative, since: prescinding from the participation—formal or otherwise—in the morally illicit intention of the principal agent, the case in question entails a proximate material cooperation in the production and manipulation of human embryos on the part of those producing or supplying them." I see this instead as immediate material cooperation with the evil of human embryo destruction.

27. For example, the Kitasato Institute of Japan has developed a vaccine for rubella that is grown in rabbit kidney cells. This product has not been approved by the FDA for use in the United States, but it is currently available in Japan. There is no strictly scientific reason why these vaccines must be grown in human cell lines.

28. A recent example of a remarkable and very important study is Jiang et al., "Pluripotency of Mesenchymal Stem Cells Derived from Adult Marrow," *Nature* 418 (2002), 41–49. This study showed that stem cells in adult bone marrow were able to form all three types of descendant cells that make up the human body. The descendant cells showed no loss of differential potential in vitro and were described by the authors as an "ideal cell source for therapy" for diseases.

29. McHugh and Callan, *Moral Theology,* nos. 1522–23.

CHAPTER 8

Stem Cells and Social Ethics

Some Catholic Contributions

LISA SOWLE CAHILL

Many theologians and philosophers, politicians and members of the public find stem cell research to be full of ethical perplexities and imponderables. In the 1960s, John Noonan, the Catholic historian and legal scholar, confronted a similar situation as the birth control debate was heading toward the crisis of *Humane Vitae*. Placing the evolution of Catholic teaching on contraception in a social context extending over time, Noonan defines "the moralist's business" as "the drawing of lines"[1]—not necessarily indelible lines, but prudential determinations of how to protect values while negotiating ambiguity, uncertainty, and conflict.

In the instance of sex, Noonan identifies the enduring values as procreation, education, life, personality, and love. In stem cell research, the two most obvious values at stake are nascent life and medical benefits. But constellations of values, the dangers to which they are subject, and specific opportunities for their defense and enhancement change over time and in relation to historical factors.[2] In assessing the ethics of embryonic stem cell research, advocates and opponents both tend to assert more clarity than the situation warrants. Drawing lines about specific ethical stances and policies will not be easy.

I will propose that embryonic stem cell research ought to be approached very cautiously and in a spirit of great moral reservation; the lines we draw are prudential and call for practical wisdom. Ultimately, I think our key concerns should be advocating for basic health care and drawing the firmest line possible on the creation or cloning of embryos for research purposes. The research use of embryos that are to be discarded in any event may be justifiable, but it is not unproblematic.

VALUES IN THE STEM CELL DEBATE

I will begin by identifying, like Noonan, five general values that we should be committed to protect in the moral debate about embryonic stem cell research. These are: the value of nascent life, the value of moral virtue or moral integrity, the value of medical benefits, the value of distributive justice or just institutions, and the value of a social ethos of generosity and solidarity. To illuminate the ways in which these values and general moral principles advancing them might be represented in detailed judgments and practices, I will draw on five principles of traditional Catholic theology (one in the case of each value). These are probabilism, cooperation, double effect, common good, and preferential option for the poor (a more recent emphasis in Catholic thought).

1. The Value of Nascent Life. This value yields a principle of respect for and protection of early human life, furthered under the traditional prohibition of directly taking innocent human life. Does the life of the embryo fall directly under this prohibition? Is the prohibition absolute in this case?

One of the crucial moral conflicts posed by the use of an embryo's stem cells for research is that the benefits it offers must be obtained at the price of its life. But the precise moral status of the embryo is a highly contentious matter. Although it is hard to deny that even a recently fertilized human egg is a living organism and a member of the human species, it is not so clear that very early human life has the full moral status of a person. Nonetheless, official Catholic teaching asserts that the earliest embryo, the blastomere, is morally the same as a person for all practical purposes. For example, referring specifically to the use of embryonic stem cells, John Paul II has ruled out "the manipulation and destruction of human embryos," on the grounds that "methods that fail to respect the dignity and value of the person must always be avoided."[3] On the other end of the spectrum are *New York Times* editorials that have referred repeatedly to the stem cell source as just "a microscopic clump of cells."[4]

Somewhere in the middle are a variety of other positions that see embryonic value as developing over time. The late Richard A. McCormick popularized the phrase "nascent human life" to refer to developing life whose value is not merely potential but that also is not fully realized individual human life until later.[5] Since at least the late 1960s, Catholic authors have proposed that the time of "individuation" and "implantation," in the second week after fertilization, offers a reasonable line defining the emergence of decisive value in embryonic life.[6] At this point, the time of possible twinning is past, the embryo implants in the uterine wall, and its chances of survival to birth increase enormously. In Great Britain and Australia, research on embryos younger than fourteen days is permitted for these reasons.

Restating this position in light of the stem cell debate, Kevin Wildes defends a "developmental school," holding that "while the early human embryo is worthy of respect, it ought not to be given personal moral status until there has been sufficient development of the embryo."[7] This development has not yet occurred, in his view, during the time at which totipotent stem cells can be extracted from the blastocyst and still develop into different human organisms. Thomas Shannon has also defended a similar view.[8] The National Bioethics Advisory Commission adopted a loose version of the developmental position in its 1999 report on the federal funding of stem cell research: "We believe that most Americans agree that human embryos should be respected as a form of human life, but that disagreement exists both about the form that such respect should take and about what level of protection is owed at different stages of embryonic development."[9] Among the "shared values" that became part of the framework of their report, the commissioners included "respecting human life at all stages of development."[10] The report, however, went on to approve the destruction of embryos for research purposes, though not their creation for research. This limit was phrased in nonabsolute terms, however, depending on the fact that "at this time" there is no "public support for creation of research embryos."[11]

In a well-nuanced discussion in the *Hastings Center Report*, Michael Meyer and Lawrence Nelson show how respect and regret can and do surround the destruction of some types of valuable entities, including cadavers, animals, and even fetuses. However, the authors ultimately must ground their view that embryos can be destroyed respectfully on the premise that their status is "weak or modest."[12] They recommend that only embryos less than fourteen days old be used, that they be used only as a last resort for important research (which might rule out embryonic stem cell research, since its advantages over adult stem cell research have not been well demonstrated), that they not be sold, and that the actions of researchers should demonstrate loss, regret, and seriousness.

A subsequent issue of the *Report* featured two letters of objection, testifying to the paradoxical, even if not incoherent, nature of a policy of "respect and destroy." Daniel Callahan calls this a mere "cosmetic ethics . . . of no value whatever to the embryos," serving "only to make the embryo donors and the researchers feel better." Cynthia Cohen suggests that "it is difficult, even across cultures, to escape the nagging realization that a human embryo is a human being in process" and that as "a potential human being, it has more than weak moral status."[13]

Although drawing lines at fertilization, fourteen days, or even birth can provide clear moral and policy guidance, this guidance might come at the price of consistency with objective reality: for instance, the contradiction in seeing the embryo as a human but "nascent" or developing entity that has significant value from fertilization but does not fall within the category "person." It is very difficult to recognize the moral relevance of embryonic development in a clear and coherent formula that can connect with meaningful moral and policy requirements prescribing and limiting the ways in which embryos can actually be treated, beyond just the attitudes accompanying destructive research.

The clarification of the value of embryonic life in the 1987 Vatican *Instruction on Respect for Human Life in Its Origin and on the Dignity of Procreation* is interesting in light of the above uncertainties. Its position on the personal status of the embryo is more nuanced than some might think. First, the opening sections of the document refer to the specific problem of infertility therapies, biomedical research, and science and technology. The concluding paragraphs refer to civil society, political authority, legislation, and the courts. The specific guidelines on treatment of the embryo are thus set in a context that envisions and takes into account the larger implications for science, society, and policy of rulings that are made. The document asserts in seemingly unqualified terms that "from the moment of conception, the life of every human being is to be respected in an absolute way."[14]

Shortly thereafter, however, the document observes that, although "the human being is to be treated as a person from the moment of conception," the presence of a human soul cannot be proven by any "experimental datum," nor has the magisterium "expressly committed itself to an affirmation of a philosophical nature."[15] I infer from this that the embryo, while not proved scientifically or philosophically to be a person, is still to be given the benefit of the doubt and treated as a person, not for special religious reasons, but because the preponderance of scientific and reasonable evidence is judged to fall on the side of personhood. (". . . The conclusions of science regarding the human embryo provide a valuable indication for discerning by the use of reason a personal presence at the moment of the first appearance of a human life: how could a human individual not be a human person?"[16])

But what if doubt in the matter is more significant than this document concedes? What if doubt as to the early embryo's full personal status is substantial? How does the ambiguity of the status of the embryo interact with other conditions and social factors that come into play in practical decision making? Probabilism, cooperation, double effect, and common good are all traditional ways of acknowledging the social and practical character of moral discernment. First, probabilism. Catholic moral tradition has not treated degrees of doubt as all having the same moral bearing on the permissibility of a given action. The larger the doubt as to the rectitude of a certain moral teaching or the facts upon which it relies, the greater the room there is for flexibility in the application of the teaching. The pertinent traditional debate was over when freedom from obedience to a doubtful law could be justified. How is it possible to move from a state of uncertainty about the application of a law to a choice about a right course of action?[17] The name most often attached to this debate is that of Alphonsus Liguori, an eighteenth-century defender of the moderate position that a solid opinion against the law is sufficient to justify freedom, even if other opinions are also probable. But the roots of this solution go back to the sixteenth-century Dominican, Bartolomeo de Medina, who wrote in his commentary on Aquinas that "'if an opinion is probable . . . it is permissible to adopt it, even if the opposite be more probable.' And if one wishes to know what constitutes a probable opinion, it is one 'stated by wise men and confirmed by good arguments.'"[18] In the view of a pre–Vatican II commentator, "Probabilism is common sense; it is a system used in practical doubt by the majority of mankind."[19]

In the case of stem cell research, it is not necessary to reject the general principle that it is wrong directly to kill an innocent human person in order to call into question the application of the law in the case of the early embryo. Yet, according to at least some interpretations of probabilism, the maxim that "a doubtful law does not oblige" is meant to cover only some types of doubt. It does not apply in cases in which, rather than there being a doubt about whether the law is meant to cover a given sort of case (a doubt of law), there is a doubt of fact about the real circumstances to which the law is to be applied, a doubt which would put natural rights in question, particularly the right to life. For example, one may not shoot at movement in the brush that might probably be a man, even if it is more probable that it is not.[20] In a lengthy article on probabilism and the early embryo, Carol Tauer has argued that probabilism does apply to doubt whether the early embryo must be treated as a person, since its status is not a matter of "fact" in the sense used by the theorists of probabilism. The status of the embryo cannot be settled as matter of empirical fact, but requires a philosophical judgment; yet all the examples of doubts of fact given by traditional moralists refer to empirically verifiable states of affairs. It would be "ludicrous," Tauer concludes, to apply the law against taking human life in such a way that "if there is the slightest chance that some type of being falls under the law, then we may not kill it."[21]

An issue that deserves perhaps further attention is the problem of weighing the probability that the principle of the inviolability of innocent human life *does* apply in the case of the early embryo, both against the certitude or risk of harm that will follow from insisting on a strict application and against indirect deleterious effects on the social ethos that could result from a lenient interpretation. In the case of stem cell research, to treat the embryo as a person is possibly to forego lifesaving benefits to undoubted persons; but to not treat it as a person may lead to the instrumentalization and commodification of life, if firm standards for the treatment of nonpersonal "nascent" life cannot be established. We shall return to these matters below, in considering the values of medical benefits, distributive justice, and solidarity.

2. The Value of Moral Virtue. This value may be translated into the principle that one ought always to act with moral integrity and never to act against one's conscience. Conscience, in turn, should be informed with reference to a reasonable interpretation of one's objective moral circumstances and the effects of one's actions on others and on society.

The value of moral integrity impinges on the stem cell debate by way of the prospect of cooperation with or complicity in the evil of destroying embryos. Whether this is a moral evil remains to be decided on the basis of a more complete discussion, but few would deny that ending the lives of embryos is at least a prima facie strike against the justifiability of stem cell research. To be involved in or to support the destruction of embryos thus poses a danger to the virtue and integrity of moral agents who otherwise would support this research. The object of moral concern here has recently been identified in the press and in policy debates as "complicity" in evil. The reason President Bush limited the cell lines to be studied with federal funding to the sixty or so existing as of August 9, 2001, was to avoid encouraging the derivation of more cell lines through the destruction of additional embryos in the future. While many have objected that this degree of availability is far too inadequate, others have criticized Bush's decision as permitting undue "complicity" in the evil of the original derivation.

What sorts of moral concerns are denoted by the term "complicity"? They fall into two essential categories: (1) the moral standing of the agent who proposes to be in some way associated with evil actions undertaken by another and (2) the effects of complicity on the social context or community in which the actions take place. In Catholic moral theology, a principle that addresses similar but not exactly identical concerns is that of cooperation.[22] Like the principle against complicity, the principle of cooperation addresses both the intention and hence moral posture of the agent as well as his or her role as a possible facilitator of evil whose actions have an unjust impact on others. One difference between cooperation and complicity is that the former judges action prospectively, while the latter also looks back to past immoral actions whose consequences one now appropriates as part of one's own action. While it is clear that participating in a proposed immoral act in order to achieve some good is likely to have the objectionable effect of facilitating or encouraging the evil, it is less clear that using the results of a past bad act to achieve some subsequent good entangles the second agent in the original evil in the same way. The past evil is over and done, and later choices cannot prevent it. Nonetheless, as debates about the moral justifiability of using the results of Nazi medical experiments have shown, complicity in past evildoing, even by those who do not approve of it, is a serious moral possibility.[23] Important to consider is the likelihood that present use of past immorality will encourage similar wrongdoing now or in the future, by seeming to give it moral, practical, and social acceptability. While one would hope that there exists a strong social consensus about the immorality and intolerability of Nazi-like experiments, the same cannot be said for the destruction of embryos in pursuit of stem cells, which is already occurring at a considerable rate in privately funded laboratories.

What light can the principle of cooperation shed on the stem cell dilemma? According to a moral theology text widely in use before Vatican II, cooperation is "concurrence with another in a sinful act," which is always wrong if it is "formal," i.e., if both agents fully intend and desire that the act occur. In cases of "material" cooperation, the cooperating agent shares in or facilitates a wrong deed, without necessarily wanting it to occur. Not only may the cooperating agent be acting under pressure, but his or her cooperation can be tied to the wrongful act by degrees: direct participation (immediate material cooperation), "secondary and subservient" participation (mediate material cooperation), and secondary participation that is also removed in time and place from the wrong act itself (remote mediate material cooperation).[24] Fairly clearly, the use of embryonic stem cells derived by others falls into the latter category, with the additional distancing factor that the objectionable act with which one "cooperates" (or better, is complicit) has already occurred.

According to Henry Davis, the author of the widely used text, cooperation of this sort can be justified if the cooperating act (here, research on the stem cells) is not in its own right sinful, and if there is a sufficient cause (here, development of therapies for disease, the prestige of cutting-edge research, and the financial gain to ensue). Of course, the "sufficiency" of the cause must be determined in relation to the weightiness of the wrongful act, which in the case at hand, returns us to the debated question of the precise status of the embryo.

Leaving the complicity question to one side for the present, as not in itself definitive of the ethics of stem cell research, we shall move on to the next point.

3. The Value of Medical Benefits. The rationale for destroying an entity that deserves, at the least, "special respect" turns on the immensity of the health benefits that supporters of the research envision for the future. The basic ethical principle of beneficence implies that we should serve the health of human beings by developing medical science and technology.

The report on the current state of the science of stem cells prepared by the National Institutes of Health (NIH) for President Bush in June 2001 describes research on stem cells as having "extraordinary promise."[25] More specifically, the report mentions the hope that cells may be replaced in "many devastating diseases," such as Parkinson's disease, diabetes, chronic heart disease, end-stage kidney disease, liver failure, and cancer. Replacement tissue may be generated to treat neurological diseases, including spinal cord injury, multiple sclerosis, Parkinson's disease, and Alzheimer's disease.[26] Yet the magnitude of these projections is mitigated by their degree of uncertainty. To date, there are no diseases for which stem cell therapy has been proven effective in humans. *Commonweal* magazine entitled an editorial "The Stem-cell Sell" and, perhaps hyperbolically, compared "scientists clamoring for federal funds" to "that quintessential American huckster, the snake oil salesman."[27] The fact is that profits as well as humanitarianism motivate stem cell science advocates, as will be discussed in the next section. Promises should be received with a healthy dose of skepticism. At the same time, it is hard to doubt that the medical advances to be achieved through such research could be significant.

Catholic moral tradition has developed tools for judging when it is justified to cause some evil in the pursuit of a good. For a utilitarian ethic, the sole moral criterion is a balance of good over evil effects, the greatest good for the greatest number. In its more subtle invocation of multiple criteria, Catholic tradition concurs in the moral value of seeking to bring about good as widely as possible, but sets limits on the means that may be used in so doing. The traditional principle through which this is accomplished is called double effect. The principle of double effect has been the subject of a huge amount of intra-Catholic debate since the 1960s because, taken together, its criteria seem to rule out good-producing actions that common sense would condone, and because the principle's constituent criteria do not, in fact, hang together all that coherently. The principle may perhaps best be understood as a practical summary of the common features of situations in which promoting good at the cost of some evil should be permitted. It is not a mathematical formula for guaranteeing moral rectitude, nor even for ruling out classes of actions beyond a shadow of a doubt. But as a practical, prudential guide to moral discernment, it is still useful.

The conditions of double effect can here be set out only briefly and with admittedly inadequate exposition and critique. According to the Jesuit medical moralist Gerald Kelly, who published a widely used commentary on the *Ethical and Religious*

Directives for Catholic Hospitals in 1957, double effect is a basic tool of moral reasoning. An action that brings about good while producing an evil effect is permitted if the following conditions are met: (1) the action, considered by itself and independently of its effects, must not be morally evil; (2) the evil effect must not be the means of producing the good effect; (3) the evil effect is sincerely not intended but merely tolerated; (4) there must be a proportionate reason for performing the action, in spite of its evil consequences.[28] The debates about this principle have centered primarily on an implied category of absolutely forbidden "intrinsically evil" acts, under the first criterion; and on the necessity of ensuring that the evil effect not be the means to the good, as required by the second criterion. Some revisers of the principle have offered the opinion that double effect's core lies in the third and fourth criteria, so that if the main object of one's intention is the accomplishment of the good effect, and if that effect is greater than the harm caused, then the act is permitted.[29]

Assuming that the "evil effect" under consideration is the death of embryos, the results of applying double effect will differ, depending on the status or approximate status assigned the embryo. Those who view it as a person would rule out research even to bring great benefits, since killing an innocent person is regarded as an "intrinsically evil act" and in violation of the first criterion of double effect. However, moving to criteria (2), (3), and (4), the research on the stem cells is not in itself the means of killing the embryo; the death of the embryo is in fact not wanted in its own right, but only as a means to a good end; and the saving of many lives could be seen as proportionate to the deaths of a more limited number of embryos. Thus, if the status of the embryo is less than clear, the force of the first criterion prohibiting intrinsically evil acts is equally in doubt, and the application of the principle follows suit. Killing a living being, perhaps even a nascent human being, is not necessarily intrinsically evil, if that being is not a person. The other criteria can be met. Thus, perhaps destroying embryos could be justified by anticipated benefits. But ambiguity about this result is commensurate with the remaining levels of uncertainty about the status of the embryo itself and the real potential of stem cell research to result in major advances. Still, the principle of double effect serves as a reminder that "the greatest good for the greatest number" is a valid but not self-sufficient moral principle. There must be limits on the means used to bring about even the best of consequences, however difficult these limits may be to set and maintain.

4. The Value of Distributive Justice. Equitable sharing in social goods, including the goods of health and health care, is a value protected under the larger value and virtue of justice. The Catholic common good tradition upholds distributive justice as a requirement of social ethics.

While commutative justice calls for fairness in the relations between individuals, for example, in undertaking and fulfilling contracts, distributive justice focuses on the community as a whole and its distributions of benefits and burdens to all through its government and institutions. Distributive justice is implied by and furthered through the principle of the common good. The common good has a long history in Catholic social ethics, from Thomas Aquinas to the modern papal social encyclicals. The common good is a concept of justice that begins from the sociality of the person and includes mutual rights and duties of all members of society; the cooperation and participation of all in the common good of society, so that all contribute to and share in society's material and social benefits; and the moral and legal responsibilities of the state, which extend beyond guaranteeing civil liberties and freedom to ensuring that the basic needs of all are met. Substantive goods in which all are entitled to share include food, shelter, education, employment, private property, and political participation. The right to own private property is limited by the rights of all to basic goods and the duties of all to contribute to the common welfare. Since the 1960s the concept of the common good has been expanded to include a global dimension. John XXIII called for international cooperation to end the threats posed by nuclear deterrence and the cold war; Paul VI called for more responsibility on the part of the industrialized nations to aid in development through trade, aid, debt relief, and investment; John Paul II has noted repeatedly how the consumerism and materialism of some countries results in economic and cultural deprivation for others, and he has called for a new spirit of solidarity to inform renewed commitment to the common good. Bioethicist Andrew Lustig applies this ethic of the common good to health care, stressing that social justice has an institutional and structural meaning. Societies and governments are under a moral requirement to mediate the claims of individuals, to advance the right to medical care, to address social inequities through institutional change, and to prioritize the dignity of the most disadvantaged in society.[30]

Although John Paul II's remarks to President Bush on the inviolability of the embryo have been quoted frequently, less attention has been given to his accompanying words on justice and the common good. Yet these have an equally important bearing on the social ethics of stem cell research. Meeting with Bush, the pope called for

> a revolution of opportunity, in which all the world's peoples actively contribute to economic prosperity and share in its fruits. This requires leadership by those nations whose religious and cultural traditions should make them most attentive to the moral dimension of the issues involved. Respect for human dignity and belief in the equal dignity of all the members of the human family demand policies aimed at enabling all peoples to have access to the means required to improve their lives. . . .[31]

Meanwhile, national and international patent law are expanding to permit highly remunerative development and marketing of discoveries in genetic science and biotechnology, including stem cell science. Investment in new scientific research in the biomedical sciences will guide such research toward the most profitable potential consumers. Those with the best ability to pay will have the most, or even exclusive, access to the benefits promised. Corporations are increasingly gaining control over research, new technologies and treatments, targeted audiences, and profits. A recent cover story in the *New York Times* "Money and Business" section featured the legal toehold of Geron Corporation in the stem cell market, while it also emphasized that Geron has as yet no real stem cell products to sell.[32] Geron controls the commercial rights to most of the stem cell lines that now exist in the United States, and it has already been involved in disputes over how widely they will be available and at what price. Dr. Thomas Okarma, Geron's president, asserts that his company "is going to dominate regenerative medicine." Although he claims not to want to impede others from doing research, he also intends to charge royalties for the development of stem cell–based therapies for diseases like diabetes. "I'm not apologetic for our intellectual property. We paid for it, we earned it and we deserve it." Vouching that he was raised a Catholic, Okarma questions why his work is being challenged by the pope. "'What is the objective of religion?' he asks. 'To make society better. Isn't that what we're trying to do here?'"[33] The distributive justice question is who belongs to the society of recipients of the improvements at which stem cell researchers aim. In a nation with forty-four million medically uninsured persons, the society of beneficiaries is not going to be inclusive. In a world in which millions of people live on less than a dollar a day and die from lack of clean drinking water, nutrition, basic health care, and diseases like malaria, tuberculosis, and AIDS, the regeneration of tissue by stem cell techniques is as exotic as it is expensive.

Before returning to that issue in the next section on social ethos, let me make a final observation on distributive justice in access to stem cell therapies. Finally, to the extent that stem cell lines are taken from embryos left over from in vitro fertilization (IVF), they are unlikely to provide the best tissue matches for a diversity of populations, even if wide access were financially available. Patients seeking treatment at infertility clinics are overwhelmingly white. Most of the cell lines were cultivated from embryos in the United States, Sweden, Israel, Singapore, and India. This provides for very limited genetic diversity in any therapies eventually developed from the cells.[34] While humans are 99 percent the same genetically, and while race is an unreliable, loose, and overlapping category on which to make human distinctions, it is still true that propensity to disease varies significantly according to ethnic and racial backgrounds. For example, those of white European ancestry are more susceptible to cystic fibrosis, light-skinned women to osteoporosis, whites more than blacks to Alzheimer's, blacks to sickle-cell anemia, and Ashkenazi Jews to Tay-Sachs disease.

5. The Value of a Social Ethos of Generosity and Solidarity. Speaking from a religiously informed standpoint and drawing on liberation theology and the Polish solidarity movement, John Paul II and others adopt the principle of a "preferential option for the poor" and see solidarity as the social virtue most important to the common good. Generosity and solidarity can help create a social ethos endorsed and shared by persons from many moral and religious traditions, for they reflect the highest ideals of humanity and social justice.

In the 1980s and 1990s, during the current pontificate, the "preferential option for the poor" became an increasingly visible part of the common good tradition, often imaged in gospel terms and associated with the virtue of "solidarity." John Paul II has been a strong critic of consumerism, materialism, and the excesses of market capitalism. In *Evangelium Vitae* (Gospel of life), he rejects "a completely individualistic concept of freedom, which ends up by becoming the freedom of 'the strong'" (no. 19), commends greater international availability of medical resources as "the sign of a growing solidarity among peoples" (no. 260), and reminds us that "it is above all 'the poor' to whom Jesus speaks in his preaching and actions" (no. 32).[35] A couple of weeks prior to his meeting with George Bush to discuss stem cell research, he had addressed a congress of Catholic doctors, urging "researchers in the biomedical sciences" to "make a generous contribution to providing humanity with better health conditions" and to cultivate "a deeper concern for your neighbor, a generous sharing of knowledge and experience and an authentic spirit of solidarity. . . ."[36]

The development of profitable medical miracles for the elite extends as an ethical problem across the spectrum of biotechnology and genetics research. The legal scholar and philosopher Margaret Jane Radin notes the growing prevalence of market rhetoric in many spheres of cultural and political life. While granting that commodification is not always inappropriate, and that commodification admits of degrees, sometimes acceptable ones, she warns that "commodification of significant aspects of personhood cannot be easily uncoupled from wrongful subordination," and that the commodification of persons and of significant personal relationships can result in "dehumanization and powerlessness."[37] A Canadian legal scholar, commenting on genetics research, notes similarly a "commercialization environment" that can be seen worldwide, but which he finds most clearly present in the United States, where individual choice and the right to buy and consume products is so key to the national culture.[38]

Instances of excessive commodification and subordination can be hard to identify with precision, and there may be legitimate disagreement in this regard.[39] Yet, to the extent that basic human goods are commodified, and that their availability depends on their purchase as commodities, some persons who need them are certain to be excluded. When embryos are destroyed and their cells sold to provide saleable research material and to promote biomedical business, commodification increases in the sphere of procreation, which should generally be an expression of sexual, parental, and familial commitments.

A social ethos of generosity and solidarity does not require that the market in biotechnology and medicine be eliminated, but that it be limited by a sense of the common good, mutual rights and duties, and the participation of all in goods to be shared. Solidarity and generosity, rather than commodification and the profit motive, seem particularly important as virtues to guide the treatment of early human life as well as of lifesaving medical measures.

CONCLUDING REFLECTIONS

A German colleague with whom I have participated in an international bioethics seminar, Dietmar Mieth, is a bioethicist at the University of Tübingen. He serves on ethics committees for UNESCO (United Nations Educational, Scientific, and Cultural Organization) and the European Parliament, and has labored for years in a religiously, ethically, culturally, and politically pluralistic environment to create an international ethos that is more protective of embryos, less willing to accept genetic interventions whose results for future generations are unknown, and less willing to sacrifice moral values for the sake of scientific advances and investments. Working in the midst of differing ethical perspectives, factual uncertainties, moral ambiguities, and policy limitations, he has come to rely on what he calls a "convergence argument": an ethical position based on several arguments, none of which is conclusive in itself, but which together "form a sort of cable made of different cords."[40]

On the ethics of stem cell research, I have woven together five cords: the special status of the embryo, moral integrity, medical benefits, distributive justice, and an ethos of generosity and solidarity, to form what is a suggestive rather than a conclusive argument. In the case of leftover IVF embryos, the fully personal status of the embryo is in question, as was demonstrated above in the discussion of the value of nascent life. I see the fact that frozen embryos have virtually no prospects for gestation and birth as a relevant consideration, since it means that their destruction deprives them, at most, of indefinite existence in an arrested state of development, and not of the development of personhood. Perhaps an analogy to brain-dead patients who become organ donors could be made. It is important to draw our firmest lines against the creation of embryos for research, since it is a proximate social danger and since it constitutes a major step toward the commodification of developing human life and of the procreative process.

The thrust of this paper has been to show that the ethics of stem cell research is a complicated and ambiguous territory that cannot be negotiated by any simply scientific or philosophical argument. The Catholic tradition does provide tools—like the principles of double effect and cooperation—to guide moral decision making in gray areas. However, in the case of stem cell research, not even traditional principles and values are able to provide incontrovertible answers. The most important role of thoughtful ethicists, Catholic and otherwise, may be to support the emergence of a different ethos, one more attuned to social interdependence and reciprocity regarding basic needs and less trusting of technological and scientific predictions and marketing propaganda. Certainly, this would mean more attention to distributive justice, and more caution in using embryos for what is claimed to be scientific necessity, at least until other avenues, such as adult stem cell research, have been explored. The centerpiece of the Catholic agenda for medical research should be basic health care and solidarity. Concerns about the embryo should be framed, as the words of John Paul II suggest, within a larger call for "respect for human dignity and belief in the equal dignity of all the members of the human family," achieved through "policies aimed at enabling all peoples to have access to the means required to improve their lives. . . ."[41]

NOTES

1. John T. Noonan, Jr., *Contraception: A History of Its Treatment by the Catholic Theologians and Canonists,* enlarged ed. (Cambridge, Mass., and London: Harvard University Press, 1986), 459.

2. Of Paul VI's reconsideration of acceptable means of controlling fertility, Noonan remarks, "The campaign of the Church to outlaw some forms of contraception must be read in conjunction with [a] doctrinal development that came eventually to allow, for example, the deliberate use of the sterile period to avoid pregnancy." Ibid., 473.

3. John Paul II, "Address to the 18th International Congress of the Transplantation Society," 29 August 2000, available at the Vatican website, www.vatican.va.

4. For example, editorial, "Sensible Rules for Stem Cell Research," *New York Times,* 25 August 2000, A2.

5. Richard A. McCormick, S.J., "Notes on Moral Theology: 1978," *Theological Studies* 40 (1979): 108–9.

6. See Joseph Donceel, S.J., "Immediate Animation and Delayed Hominization," *Theological Studies* 31 (1970): 79–80; and Thomas A. Shannon, "Human Embryonic Stem Cell Therapy," *Theological Studies* 62 (2001): 811–24. For an overview of some other literature on the topic, see Lisa Sowle Cahill, "The Embryo and the Fetus: New Moral Contexts," *Theological Studies* 54 (1993): 124–42.

7. Kevin Wm. Wildes, S.J., "The Stem Cell Report," *America,* 16 October 1999, 14.

8. See Shannon, "Human Embryonic Stem Cell Therapy," 816–19.

9. National Bioethics Advisory Commission, *Ethical Issues in Human Stem Cell Research,* vol. 1, *Report and Recommendations* (Rockville, Md.: National Bioethics Advisory Commission, 1999).

10. Ibid., 4.

11. Ibid., 55.

12. Michael J. Meyer and Lawrence J. Nelson, "Respecting What We Destroy: Reflections on Human Embryo Research," *Hastings Center Report* 31, no. 1 (2001): 18.

13. Daniel Callahan and Cynthia B. Cohen, "Letters: Human Embryo Research: Respecting What We Destroy?" *Hastings Center Report* 31, no. 4 (2001): 4 and 5, respectively.

14. Congregation for the Doctrine of the Faith, *Respect for Human Life in Its Origin and on the Dignity of Procreation: Replies to Certain Questions of the Day* (1997), "Introduction," 5; see *New York Times,* 11 March 1987.

15. Congregation for the Doctrine of the Faith, *Respect for Human Life in Its Origin . . . ,* part I.1.

16. Ibid.

17. Charles E. Curran, *The Catholic Moral Tradition Today: A Synthesis* (Washington, D.C.: Georgetown University Press, 1999), 63.

18. Quoted by John Mahoney, *The Making of Moral Theology: A Study of the Roman Catholic Tradition* (Oxford: Clarendon Press, 1987), 136.

19. Henry Davis, S.J., *Moral and Pastoral Theology,* vol. 1, *Human Acts, Law, Sin, Virtue* (New York: Sheed and Ward, 1946), 93.

20. Ibid., 99.

21. Carol A. Tauer, "Probabilism and the Moral Status of the Early Embryo," in *Abortion and Catholicism: The American Debate,* ed. Patricia Beattie Jung and Thomas A. Shannon (New York: Crossroad, 1988), 78.

22. See Shannon, "Human Embryonic Stem Cell Therapy," 814–21; James F. Keenan, S.J., "Prophylactics, Toleration and Cooperation: Contemporary Problems and Traditional Principles," *International Philosophical Quarterly* 29 (1989): 205–20; James F. Keenan, S.J., "Institutional Cooperation and the Ethical and Religious Directives," *Linacre Quarterly* 63, no. 3 (1997): 44–67; and James F. Keenan, S.J., and Thomas R. Kopfensteiner, "The Principle of Cooperation: Theologians Explain Material and Formal Cooperation," *Health Progress* (April 1995): 23–27.

23. See Robert L. Berger, "Nazi Science—The Dachau Hypothermia Experiments," in *Medicine, Ethics, and the Third Reich: Historical and Contemporary Issues,* ed. John Michalczyk (Kansas City, Mo.: Sheed and Ward, 1994), 87–105.

24. Davis, *Moral and Pastoral Theology,* 1:341–42. See also Thomas J. O'Donnell, *Medicine and Christian Morality* (New York: Alba House, 1976), 31–40; Benedict M. Ashley, O.P, and Kevin D. O'Rourke, O.P., *Health Care Ethics: A Theological Analysis* (St. Louis, Mo.: The Catholic Health Association, 1978), 197–99; and Russell E. Smith, "Formal and Material Cooperation," *Ethics and Medics* 20, no. 6 (1995): 1–2.

25. National Institutes of Health, *Stem Cells: Scientific Progress and Future Research Directions* (Bethesda, Md.: National Institutes of Health, 2001), i. Available on the NIH website, www.nih.gov/news/stemcell/sireport.htm.

26. Ibid., "Executive Summary," 4.

27. Editorial, "The Stem-Cell Sell," *Commonweal,* 17 August 2001, 5.

28. Gerald Kelly, S.J., *Medico-Moral Problems* (St. Louis, Mo.: Catholic Hospital Association, 1957), 13–14. See also Ashley and O'Rourke, *Health Care Ethics,* 194–97.

29. For some of the debates, see Charles A. Curran and Richard A. McCormick, S.J., eds., *Moral Norms and Catholic Tradition,* Readings in Moral Theology, no. 1 (New York: Paulist Press, 1979).

30. B. Andrew Lustig, "The Common Good in a Secular Society: The Relevance of a Roman Catholic Notion to the Healthcare Allocation Debate," *Journal of Medicine and Philosophy* 18 (1993): 569–87.

31. John Paul II, "Remarks to President Bush on Stem Cell Research," *National Catholic Bioethics Quarterly* 1 (2001): 617–18; summarized and cited in "Remarks by John Paul, Rome, July 23, 2001," *New York Times,* 24 July 2001, A8.

32. Andrew Pollack, "The Promise in Selling Stem Cells," *New York Times,* 26 August 2001, sec. 3, p. 1.

33. Ibid., 11.

34. John Entine and Sally Satel, "Race Belongs in the Stem Cell Debate," *Washington Post,* 9 September 2001, B1, B6.

35. John Paul II, *The Gospel of Life* (Boston: Pauline Books, 1995).

36. "Holy Father to Catholic Doctors Congress," 7 July 2000, at the Vatican website, www.vatican.va.

37. Margaret Jane Radin, *Contested Commodities* (Cambridge, Mass.: Harvard University Press, 1996), 163, 82.

38. Timothy Caulfield, "Regulating the Commercialization of Human Genetics: Can We Address the Big Concerns?" in *Genetic Information,* ed. Ruth F. Chadwick and Alison K. Thompson (New York: Kluwer Academic/ Plenum Publishing, 1999), 153.

39. Margaret Jane Radin, "Response: Persistent Perplexities," *Kennedy Institute of Ethics Journal* 11 (2001): 305–15.

40. Dietmar Mieth, "The Ethics of Gene Therapy: The German Debate," paper prepared for a meeting of the Genetics, Theology, and Ethics Group, Boston College, October 1999. This was a five-year international study group sponsored by the Porticus Foundation.

41. John Paul II, "Remarks to President Bush," 618.

CHAPTER 9

The Ethics and Policy of Embryonic Stem Cell Research

A Catholic Perspective

RICHARD M. DOERFLINGER

Proponents of embryonic stem cell research sometimes complain that religion in general, and Catholicism in particular, have become serious obstacles to what they see as progress. For example, in his new book on the embryo research debate, Professor Ronald Green—formerly vice-chair for ethics of the Human Embryo Research Panel at the National Institutes of Health (NIH)—writes that the Catholic bishops conference of the United States has been "the most active and effective center of opposition to all facets of human embryo research" and now to embryonic stem cell research.[1] Here and elsewhere, the view has been expressed that people motivated by religious belief are rather more influential in this debate than they have any right to be. The implication is that the churches, especially the Catholic Church, are trying to "impose their beliefs" about embryo research on a morally diverse society.

However, in all its many dozens of pages of testimony, public comments, and other communications to legislators and government advisory boards on this matter, the Catholic bishops conference has never asked these bodies to take actions that presume the truth of Christian revelation or specifically Catholic teaching. Catholicism has a special contribution to make to this debate—but perhaps not in the way its opponents assume.

In the first part of this paper, I wish to review some propositions on which broad consensus can be reached aside from influence by specifically Catholic views. By "broad consensus," I do not mean that there is universal consensus, but that these propositions are widely acknowledged by scientists, government bodies like the NIH, and federal advisory bodies that vigorously support destructive human embryo research. I will argue that a strong case can be made against pursuing embryonic stem cell research based solely on such broadly accepted propositions. In the second part, I address the question: If such a case can be made on nonreligious grounds, what then is the distinctive contribution of religion in general and Catholicism in particular to this debate?

A SECULAR CONSENSUS

Here I will present five propositions that, while not beyond all argument, are agreed to by many of the Catholic Church's most vigorous critics in the moral debate.

1. The human embryo, at the blastocyst stage, is a developing human life.

Four major advisory groups recommending federal policies on human embryo research over the past twenty-three years have agreed on this. For example, an Ethics Advisory Board to the Department of Health, Education, and Welfare concluded in 1979 that the early human embryo deserves "profound respect" as a form of developing human life, though not necessarily "the full legal and moral rights attributed to persons."[2] The NIH Human Embryo Research Panel agreed that "the preimplantation human embryo warrants serious moral consideration as a developing form of human life."[3] In 1999, the National Bioethics Advisory Commission (NBAC) cited broad agreement that, in our society, "human embryos deserve respect as a form of human life."[4] And in 2002, a committee of the National Academy of Sciences acknowledged that "from fertilization" the embryo is "a developing human."[5] All these bodies wanted to authorize at least some research involving destruction of human embryos. They all nonetheless conceded this point, because it fits best with the evidence.

At one time, some biologists (and some Catholic thinkers relying on their theories) believed there was a qualitative difference between the embryo less than fourteen days old and all subsequent stages of development.[6] The early embryo was dubbed a "pre-embryo" by some textbooks.[7]

Today, however, this approach is being abandoned. Some textbooks that once used "pre-embryo" have quietly dropped the term from new editions, now describing the newly fertilized zygote simply as an embryo and, therefore, as the beginning of a new human individual.[8] Other experts openly dismiss the term "pre-embryo" as "discarded" and "inaccurate."[9] Professor Lee Silver, a staunch proponent of embryo research and human cloning, has given perhaps the most startling testimony on this point:

> I'll let you in on a secret. The term pre-embryo has been embraced wholeheartedly by IVF [in vitro fertilization] practitioners for reasons that are political, not scientific. The new term is used to provide the illusion that there is something profoundly different between what we nonmedical biologists still call a six-day-old embryo and what we and everyone else call a sixteen-day-old embryo.
>
> The term pre-embryo is useful in the political arena—where decisions are made about whether to allow early embryo (now called pre-embryo) experimentation—as well as in the confines of a doctor's office, where it can be used to

> allay moral concerns that might be expressed by IVF patients. "Don't worry," a doctor might say, "it's only pre-embryos that we're manipulating or freezing. They won't turn into real human embryos until after we've put them back into your body."[10]

The early embryo was once dismissed as a mass of interchangeable and undifferentiated cells—capable at any point of splitting into two or more embryos (hence without inherent individuality)—and largely formless until the appearance of the "primitive streak" (hence without spatial orientation) at around fourteen days. However, it now seems that an embryo's potential for spontaneous "twinning" is established very early, perhaps by factors determining the thickness of the zona pellucida—so that the vast majority of embryos, from the outset, do not have the active potential to produce twins spontaneously.[11] Moreover, it now seems that the mammalian embryo's spatial orientation is largely determined during the first week of development, perhaps by signals in the outer cell wall or trophoblast.[12] The most recent findings, based on mouse embryos, even suggest that "two axes of the blastocyst become specified in the single-cell embryo." That is, the embryo's axes determining right and left, up and down, are determined by the point where a sperm first penetrates the egg.[13]

Differentiation into cells with different roles and functions also begins with the very first cell division of the early mammalian embryo. These first two cells already have different roles in embryonic development, with one largely devoted to making the embryo proper and the other to developing the support structures (placenta, etc.) needed for long-term survival.[14] Further development proceeds from this first differentiation along a definite plan, with one of the two cells dividing first, in accord with its distinct function, so that the embryo develops three cells, then four, then eight, and so on. At each stage this is no mere colony or aggregate of cells—much less a mere envelope full of genes—but an integrated, developing organism of our species.

So radically different are our new findings about the embryo that the major science journal *Nature* notes that they were "heresy" only a few years ago. Human and other mammalian embryos were once thought to become organized and give their constituent cells definite fates only after implantation in the womb; now it is found that the embryo begins differentiation and develops a "top-bottom axis" guiding future development almost immediately after conception. The journal notes that, from now on, "developmental biologists will no longer dismiss early mammalian embryos as featureless bundles of cells."[15]

New discoveries suggesting the possibility of human cloning may further adjust our perceptions of the embryo. It may be that each cell of a very early embryo can artificially be *induced* to split off and produce a new, genetically identical embryo—but it seems that, with some additional manipulation, so can each cell of our adult bodies. And if the embryo's cells are "undifferentiated" enough not to be permanently committed to any particular type of cell, it increasingly appears that the same is true of our adult stem cells—and perhaps even of our completely differentiated body cells—under appropriate conditions.[16] In short, if an organism's ability to "twin" or its cells' ability to dedifferentiate is an indication that the organism is not a human individual, then it seems none of us are human individuals.

Some ethicists have seized upon the fact that the destruction of an embryo can produce relatively undifferentiated stem cells in culture. They assume this means the cells were interchangeable and without direction when they were part of the intact embryo. This is simply a fallacy. The fact is that these cells respond promptly to changes in their environment. If they are ripped out of the developmental context in which they played a role within an organized whole, they can (as it were) lose their sense of direction and revert to a less specialized state. As the National Institutes of Health's report on stem cell research notes, however, it is probably misleading to say that pluripotent stem cells exist in the living embryo. Rather, pluripotent stem cells "develop in tissue culture *after* they are derived from the inner cell mass of the early embryo. . . ."[17]

This ability to revert to a more versatile state seems to be an effective survival mechanism, an important reason why human embryos are so resilient in the face of threats from their environment. They can respond to a loss of one or more cells by compensating, because their progressive differentiation is not yet "locked in stone" even to the (limited) extent that adult cells are.

In short, while it makes no sense to say that I was once a body cell, or a sperm, or an egg, it makes all the sense in the world to say that I was once an embryo. For the embryo is the first stage of my life history, the beginning of my continuous development as a human organism. This claim makes the same kind of sense as the claim that I was once a newborn, although I do not have any recollection of cognitive or specifically human "experiences" during either stage of my life. This is not automatically to say that the embryo has the moral status of a person. However, one need not make claims about personhood to understand a moral obligation to respect and protect developing human life. Moreover, by secular reasoning one can make a case in favor of treating every fellow member of the human species *as* a person, particularly since any *other* standard—assigning personhood on the basis of appearance, cognitive ability, etc.—will exclude many more people from personhood than just the embryo.

2. Pursuing medical treatments through embryonic stem cell research requires us to condone destroying this human life.

This proposition would be too obvious to mention if not for the fact that some have strained to deny it. Even the National Bioethics Advisory Commission conceded that promoting embryonic stem cell research implicates us in the destruction of embryos.[18] The Clinton administration's guidelines for such research demanded that, as part of informed consent, parents be told that their embryos will not survive the cell-harvesting process.[19] Yet some very creative theories have been proposed to obscure this fact. One ethicist, for example, has suggested that, since an embryo is essentially a package for genetic material and its distinctive genome survives the harvesting process, those who reduce the embryo to a cell culture are not destroying anything of importance but actually helping what is unique about the embryo to "live on and on."[20] This is like saying that if we dismantle a building and preserve the resulting pile of bricks, we are not really "destroying" the building.

Others have suggested that, since the "spare" embryos from fertility clinics would have been discarded anyway, researchers are not causing any net loss of life—they are influencing only how the embryo dies rather than whether it dies.[21] One is tempted to observe that this is true of any and all killing of mortal creatures. Such a justification could certainly have horrendous implications for lethal experimentation on terminally ill patients or death-row prisoners. In the realm of fetal research, federally funded researchers are barred by law from using the "will die soon anyway" defense to do harmful research on the unborn child intended for abortion (or the child dying outside the womb from an abortion).[22] In any case, the Clinton guidelines were not restricted to embryos slated for discarding; they extended to any embryo deemed "in excess of clinical need," which only means that the parents do not need that embryo to reproduce at the present time. Many of these embryos are kept in frozen storage and eventually transferred to a womb later (if they are not requisitioned for destructive research first).

Still others have argued that, by funding embryonic stem cell research, the government is not complicit in any destruction of embryos because the research only occurs after the embryos are destroyed. Yet Congress, since 1996, has banned federal funding of any research in which embryos are harmed or destroyed, and it is difficult to see how embryo destruction is anything but an integral and essential first step in any embryonic stem cell research project. The Clinton administration's argument that such destruction and the use of the resulting cells were completely separate activities was criticized as hypocritical and evasive even by supporters of federal funding.[23] By offering funds for research projects whose very existence relied on the destruction of embryos, the administration was, of course, knowingly encouraging such destruction to be done.

The policy articulated by President Bush on August 9, 2001, is a more subtle matter. The President's stated goal was to promote the possible benefits of embryonic stem cell research without encouraging future destruction of human embryos. Therefore, he said, federal funds would only support research using cell lines already created by destroying embryos in the past.

Setting aside the complex question of complicity after the fact, one problem here is that the policy serves two incompatible goals: funding embryonic stem cell research in the hope that it will lead to treatments and discouraging future destruction of embryos to obtain their stem cells. For, the limited number of cell lines that the Bush administration approved for federally funded research is, at best, adequate only for basic research to determine the most promising avenues for further exploration. The currently eligible cell lines are not only of insufficient volume for treatments, but they also have inadequate genetic diversity to treat most of the patients who may want cell implants; and they are grown in cultures of mouse feeder cells, which could make them inappropriate for human transplantation.[24] To solve the problem of tissue rejection, researchers will want to develop many thousands of cell lines with different genetic profiles—or develop embryo cloning, to create and destroy embryos that are a genetic "match" to each individual patient. So any promising results from federally funded research on these few cell lines can only encourage more expansive research in the private sector—research that will not respect President Bush's limits. According to some administration officials, that is precisely the intent of the policy.[25]

In short, pursuing embryonic stem cell research implicates us in justifying or sanitizing the (past) destruction of embryos to obtain the cells, and it sets the stage for more embryo destruction if the research turns out to be anywhere near as promising as its proponents claim.

3. A moral presumption against the taking of human life requires us to treat stem cell research requiring embryo destruction as a last resort, to be pursued only if medical progress cannot be achieved in other ways.

This, of course, was a key conclusion of the National Bioethics Advisory Commission's report. Having said that the embryo deserves respect as a form of human life, the commission concluded: "In our judgment, the derivation of stem cells from embryos remaining following infertility treatments is justifiable only if no less morally problematic alternatives are available for advancing the research."[26]

Many researchers, while claiming to accept this judgment, nonetheless insist that research using both embryonic and nonembryonic stem cells should be fully funded now to determine which source is best for various functions. But this approach simply reduces "respect" for the embryo to nothing at all. For that is the approach one would take if there were no moral problem whatever—if the only factor determining our research priorities were relative efficiency at achieving certain goals. "Respect" must mean, at a minimum, that we are willing to give up some ease and efficiency in order to obey important moral norms instead of transgressing them.

4. Adult stem cells and other alternatives are much more promising than once thought, offering many of the benefits once thought to be achievable only with embryonic cells.

This is apparent from the advances reported in this volume by Dr. Ira Black and others—advances that have been shamefully neglected and even suppressed by political groups pressing for embryonic stem cell research. In April 2000, for example, the Christopher Reeve Paralysis Foundation testified to a Senate subcommittee that adult stem cells are no substitute for embryonic cells because they cannot be "pluripotent" but are confined to a narrow range of specialization.[27] Yet a few weeks after that hearing, Dr. Black and other researchers funded by the NIH and the Christopher Reeve Paralysis Foundation published a study indicating that adult bone marrow stem cells "may constitute an abundant and accessible cellular reservoir for the treatment of a variety of neurologic diseases." The first sentence of the published study states: "Pluripotent stem cells have been detected in multiple tissues in the adult, participating in normal replacement and repair, while undergoing self-renewal."[28] The authors cite eleven other studies in support of this observation. Their article was received for publication in March 2000, weeks before the group funding their work testified that such findings did not exist.

One author of that study, Dr. Darwin Prockop, recently gave his own overview of advances in the use of adult stem cells:

> More than 20 years ago, Friedenstein and then others grew adult stem cells from bone marrow called mesenchymal stem cells or marrow stronal cells (MSCs). MSCs differentiate into bone, cartilage, fat, muscle, and early progenitors of neural cells. Human MSCs can be expanded up to a billionfold in culture in about 8 weeks. Preliminary but promising results have appeared in the use of MSCs in animal models for parkinsonism, spinal cord defects, bone diseases, and heart defects. Also, several clinical trials are in progress. In addition, there are promising results with other adult stem cells that perhaps we may yet learn how to grow effectively.[29]

The ability of bone marrow stem cells to form neurons and a variety of other useful cells has since been confirmed by several other research teams, leading some to speak of an "ultimate stem cell" in bone marrow that can become virtually any cell.[30]

Many examples of such progress abound, including examples of real treatments that are already curing patients, not simply speculations based on laboratory tests or animal trials. The field of "tissue engineering" using adult cells has exploded as researchers move toward rebuilding ears, tracheas, and even hearts.[31] Adult stem cells have successfully treated hundreds of thousands of patients with cancer and leukemia; they have repaired damaged corneas, restoring sight to people who were legally blind; they have healed broken bones and torn cartilage in clinical trials; they are being used to help regenerate heart tissue damaged by cardiac arrest.[32] Adult bone marrow stem cells were responsible for the first completely successful trial of human gene therapy, helping children with severe combined immunodeficiency disease to recover an immune system and safely leave their sterile environment for the first time.[33] Adult cells from a young paraplegic woman's own immune system, injected into the site of her spinal cord injury, have apparently cured her incontinence and enabled her to move her toes and legs for the first time—"generating hope for those with spinal-cord injuries around the world," as one news report observes.[34] Finally, a Canadian team has used adult pancreatic islet cells from cadavers to reverse juvenile diabetes. At the annual meeting of the American Diabetes Association in June 2001, researchers announced that all patients benefited from the transplants, and nine have remained "insulin free" for a median period of eight months—with some patients requiring no injections for up to two years.[35] Further human trials have proceeded at several sites in the United States, and "most centers have met with tremendous success."[36]

According to a recent report in *Ob. Gyn. News,* umbilical cord blood stem cells may be even more promising. "The stem cells likely to yield the quickest, least expensive, and largest clinical benefit are readily available and present no ethical dilemma. They are umbilical cord blood stem cells." These cells, already being used to treat dozens of conditions, offer "results, not hype," by contrast with the largely unrealized promise of embryonic stem cells. Even as debate rages over embryonic cells, the article notes, umbilical cord blood cells "have been steadily delivering verifiable clinical results." Used now in leukemia and other cancers, they also hold great promise for treating immune disorders, diabetes, sickle cell anemia, and other conditions.[37] In animal trials these cells have even been used to improve brain function after strokes.[38]

The NIH's report on stem cells, though firmly biased in favor of embryonic stem cell research, noted "a flurry of new information" about adult stem cells and concluded:

> Today, there is new evidence that stem cells are present in far more tissues and organs than once thought and that these cells are capable of developing into more kinds of cells than previously imagined. Efforts are now underway to harness stem cells and to take advantage of this new found capability, with the goal of devising new and more effective treatments for a host of diseases and disabilities. What lies ahead for the use of adult stem cells is unknown, but it is certain that there are many research questions to be answered and that these answers hold great promise for the future.[39]

5. There are more drawbacks and obstacles to the safe and effective clinical use of embryonic stem cells than once thought.

In a recent article in the journal *Stem Cells,* James Thomson and his colleagues at the University of Wisconsin review five challenges or obstacles to the clinical use of these cells:[40]

First, we do not know how to reliably direct these multipotent cell lines to narrow down to the kind of cell type needed and to stay that way. Even when specific growth factors are added to direct their development, they keep producing a wide variety of cells. Safe transplantation will require absolutely pure cultures of one kind of cell.

Second, when a particular cell type is produced, one must show that it will integrate into existing tissue and function in a normal physiological way. (While news media hailed the use of embryonic stem cells to treat diabetic mice in April 2001, for example, the cells in fact produced only 2 percent of the insulin needed for effective treatment and did not reverse diabetes.)[41]

Third, one must demonstrate clinical effectiveness in rodents and larger animals. Very little work has been done as yet in primates, for example.

Fourth, "the possibility arises that transplantation of differentiated human ES [embryonic stem] cell derivatives into human recipients may result in the formation of ES cell-derived tumors."[42] This problem is also discussed in the NIH's new report: "The potential disadvantages of the use of human ES cells for transplant therapy include the propensity of undifferentiated ES cells to induce the formation of tumors (teratomas)."[43] The "proliferative capacity" and tendency toward "unregulated growth"of these cells is therefore as much a danger as an advantage. Ethicist Glenn McGee has stated the problem with characteristic vividness: "The emerging truth in the lab is that pluripotent stem cells are hard to rein in. The potential that they would explode into a cancerous mass after a stem cell transplant might turn out to be the Pandora's box of stem cell research."[44]

By contrast, while nonembryonic stem cells seem harder to direct to form tissues of different categories, they seem much more docile to their environment. Upon reaching a particular kind of tissue, they receive signals as to the kind of tissue needed and produce only that tissue. They may be "easier to manage," and therefore far safer for clinical use in humans, than embryonic cells.[45]

Fifth, one must prevent rejection of these cells as foreign tissue. This may require genetic manipulation, in an effort to create a "universal donor" cell line, or the development of vast numbers of genetically diverse cell lines, or the development of human embryo cloning. In short, solving this problem may require us to create many new problems. It is a dilemma that will not exist at all if a patient's own stem cells can be directed to help heal and regenerate damaged tissues more effectively.

6. Because propositions 4 and 5 are true, one cannot safely predict that embryonic stem cells will offer clinical benefits that cannot be achieved in less problematic ways.

This was an underreported conclusion of the NIH's report. The report deemed any therapies based on embryonic stem cells to be "hypothetical and highly experimental" and said that it could not be determined at this time whether these cells will have any advantages over the less morally problematic alternatives.[46]

The only logical conclusion to be drawn from these propositions is that, particularly with public resources requisitioned from all American taxpayers, it would be irresponsible to pursue embryonic stem cell research in humans at this time. Rather, tax dollars—drawn in a great many cases from Americans with serious moral objections to the destruction of human life for research purposes—should aggressively pursue the morally noncontroversial alternatives, to see whether there is any need to discuss overriding the moral presumption established above. Certainly this approach would have broad support in public opinion. In a poll conducted by International Communications Research in June 2001, 70 percent of Americans opposed funding stem cell research that requires destroying human embryos—and 67 percent supported a policy of funding only nonembryonic stem cell research at present, to see if there is any actual need for embryo destruction to advance medical progress.[47]

This, of course, is not the approach taken by many scientists and advisory bodies. I believe that is because they are bringing into the equation some uncritically accepted axioms of their own—axioms that, for those who hold them, can fulfill the role of religious dogmas. These axioms include the materialist assumption about the nature and purpose of human life (that is, there is nothing special about the nature and nothing coherent or universal about the meaning of such life); the assumed cogency and validity of the utilitarian model for moral reflection (that is, that the end can justify the means); and the technological imperative (if we can do it, we probably should do it).

THE CATHOLIC CONTRIBUTION

This is where the Catholic Church and other religious bodies can make a particularly important contribution. They can offer a corrective to the nonfactual axioms or assumptions recounted above, thus allowing the (not solely religious) ethical case against destructive embryonic stem cell research to be assessed and appreciated on its own merits.

Some specific elements of the Catholic contribution would include first, the realism of the Catholic tradition: We live in an objective world, created without our help and filled with facts waiting to be found, some of which have great significance for our moral deliberations. It has been said that this faith-based realist metaphysics of our tradition was the main reason for the rise of modern science in Western civilization. Its significance for this debate is that it points us back to the facts recounted in the first part of this paper and insists that we take those facts very seriously. This is exactly the aspect of Catholic tradition that most frustrates Professor Green, whose objection to religious influence was noted at the beginning of this paper. His own approach, which became the basis for the conclusions of the NIH Human Embryo Research Panel in 1994, is that there are no realities "out there" in human beings that require us to respect *anyone* as a person. It is the task of the educated and articulate members of society to decide which qualities in others are morally relevant, based on their own enlightened self-interest. If we deny "personhood" to too many people, we may risk denying it to ourselves or others we care about; if we bestow it on too many people, we may be depriving ourselves of the benefits of lethal human experimentation on those people.[48]

Professor Green calls this deconstructionist approach to human dignity a "Copernican revolution" in medical ethics. It is just the opposite. Copernicus, a devout Catholic, no doubt felt he was striking a blow against human presumption by showing that we are not at the center of our universe. Green's revolution would bring back human presumption by denying any objective reality in human beings that could limit our ability to treat them as means to our ends. He objects to religion because it insists on an outside reality.

A second but closely related point is that Catholics and other believers realize that, just as we did not create the world, we do not control it. We are not gods, and we do not direct divine providence. Grandiose plans to ignore moral claims here and now in order to build an ideal future run up against the basic humility that characterizes the person of faith. We cannot claim the right to set aside the value of other people's lives to achieve something greater, because our first responsibility is to care for and respect those whom God has placed in our path. In other words, outright utilitarianism is not an option for the Catholic.

A utilitarian approach to ethics is, of course, popular among many scientists and others who propose embryonic stem cell research. When asked once whether the NIH Human Embryo Research Panel should base its conclusions on "the end justifies the means" reasoning, Professor Green quoted the man known as the father of situation ethics, Joseph Fletcher: "If the end doesn't justify the means, what does?"[49] He did not mention that Fletcher in turn cites Nikolai Lenin as the original author of this saying.[50] Lenin reportedly used it to justify the killing of countless men, women, and children to advance his revolution. History provides us with little reason to favor the application of utilitarian thinking to questions about human life. Making moral judgments solely on the basis of consequences can have bad consequences.

Third, in any debate about taking human life, believers see the stakes as being much higher—in fact as being of eternal consequence. This means that making a wrong moral decision could have eternal consequences for those of us who make it. It also means that, if the human embryo *is* a person with an immortal soul, the case is even stronger for holding that he or she is absolutely inviolable from direct attack.

Our humility prevents us from making absolute statements about whether every embryo, from the first moment of conception, has such a soul—for this is in God's hands. The closest the Catholic Church has come to doing so is to take seriously the embryological evidence on the continuity of human development from the very beginning, and to ask: How can a living individual of the human species not be a human person? Certainly we cannot accept a notion of personhood that depends on inherently changeable qualities such as possession of cognitive abilities or a certain IQ. In any case, the fact that this is a human life, at least one that God is watching over and preparing for an immortal soul, is reason enough for our respect—and the fact that this being may well already have an immortal soul makes it quite impossible to argue that we should take our chances and destroy it for the common good. To act in that way would be to show ourselves willing to commit homicide, if it turns out that our theory about delayed personhood was wrong.[51]

Fourth, our social teaching, based in faith, provides guidance on how to deal with societal questions about the common good. Our leading principles—admittedly, sometimes in creative tension with each other—are human solidarity and preferential option for the poor and vulnerable. We cannot leave anyone behind, cannot discriminate against anyone in our zeal to pursue human progress, because each is equally a child of God. In fact, if we are to give different treatment to the poor, the despised, the marginalized, it can only be to provide extra care and concern because this is where it is most needed. Patients desperately hoping for a cure obviously deserve our care and respect—but there is someone in the situation whose plight is even more desperate, and that is the individual in danger of being killed for the sake of people deemed more worthy.

Fifth, because we see faithfulness to God as our highest responsibility, we will be particularly concerned to defend rights of conscience. This arises with special urgency when government considers forcing all Americans to pay for activities that many of them find morally abhorrent. The irreducible pluralism of American society sometimes is used to argue against prohibiting immoral or controversial activities conducted by private parties; but that pluralism also argues against active government involvement in or support for activity to which many Americans have conscientious objections. The moral objections of millions of Catholics and others to treatments based on the destruction of human lives must be taken into account, not least because it would be useless and unjust to direct society's common resources to treatments that a great many may not be able to use in good conscience.

I hasten to add that these concerns are not peculiar to Catholics, and some of them can be and have been forcefully raised by non-Christians. But their centrality to Catholic moral reflection does help to explain why the church has been in the forefront of efforts to keep medical progress from relying on the destruction of human embryos.

A POSTSCRIPT ON CLONING

Similar concerns, both secular and religious, are already coming into play in the policy debate on human cloning. As mentioned above, that issue has links to the stem cell debate because it is one way scientists may want to solve the problem of tissue rejection—at the cost of making the creation and destruction of countless human embryos into an ongoing and necessary part of medical practice. To an even greater extent than in the stem cell debate, opposition to human embryo cloning may cut across the usual political and religious lines, as pro-life Catholics join with Green Party liberals concerned about genetic manipulation of human beings in order to resist a biotechnology industry that seems to recognize no absolute moral limits.

While the cloning debate is about unfit ways to create human life, the issue of destroying life will be central to that debate, for two reasons. First, even efforts to produce a live-born baby are likely to create an enormous number of miscarriages, stillbirths, and postnatal deaths, given the wastefulness and violence of the cloning procedure. Second, the solution proposed by some is to ban only what they call "reproductive" cloning: that is, to ban the transfer of embryos into the womb for purposes of pregnancy and live birth. Thus, use of the cloning procedure to mass-produce human embryos would be allowed, but government would mandate that all these embryos be destroyed instead of being allowed to survive. U.S. law, for the first time, would define a class of developing human beings it is a crime *not* to kill.

In congressional hearings and other debates, the Catholic Church has been gratified to find allies in favor of a comprehensive ban on human cloning among religious groups and secular experts who disagree with it on issues such as abortion. Some of these allies agree that the human embryo is, at a minimum, a being of inherent value and dignity that should not be cavalierly treated as research material to be created and destroyed at will. Once again, in congressional debate, proponents of cloning have accused Congress of becoming a "house of cardinals" celebrating a "papal event," because there are convergences between Catholic thought and the chief arguments used to justify a cloning ban.[52] And once again, this convergence is actually due not to some kind of Catholic takeover of public policy but to the fact that Catholic reflection on these matters is not so peculiar or so out of the mainstream as some assume. But that, perhaps, is a topic worth exploring in future symposia.

NOTES

1. Ronald Green, *The Human Embryo Research Debates: Bioethics in the Vortex of Controversy* (New York: Oxford University Press, 2001), 158. Green elsewhere concedes that the Catholic Church does not reject "all facets" of embryo research, but rather "opposes nontherapeutic manipulations of the human embryo, that is, manipulations not of direct benefit for the embryo under study." Ibid., 18.

2. "Report of the Ethics Advisory Board," *Federal Register* 44 (18 June 1979): 35033–58 at 35056.

3. National Institutes of Health (NIH), *Report of the Human Embryo Research Panel* (September 1994), 2.

4. National Bioethics Advisory Commission (NBAC), *Ethical Issues in Human Stem Cell Research,* 3 vols. (Rockville, Md.: September 1999), 1:ii; cf. 2.

5. National Academy of Sciences (NAS), *Scientific and Medical Aspects of Human Reproductive Cloning* (Washington, D.C.: National Academy Press, 2002), E-5.

6. See Thomas Shannon and Allan B. Wolter, "Reflections on the Moral Status of the Pre-Embryo," *Theological Studies* 51 (1990): 603–26.

7. See K. L. Moore and T.V. Persaud, *The Developing Human,* 5th ed. (Philadelphia: W. B. Saunders, 1993), 37.

8. See K. L. Moore and T.V. Persaud, *The Developing Human,* 6th ed. (Philadelphia: W. B. Saunders, 1998), 2. As another respected textbook notes: "Almost all higher animals start their lives from a single cell, the fertilized ovum (zygote). . . . The time of fertilization represents the starting point in the life history, or ontogeny, of the individual." B. M. Carlson, *Patten's Foundations of Embryology,* 6th ed. (New York: McGraw-Hill, 1996), 3.

9. Two leading experts now list "pre-embryo" among "discarded and replaced terms" in modern embryology, saying that it is "ill-defined and inaccurate." R. O'Rahilly and F. Müller, *Human Embryology & Teratology,* 2nd ed. (New York: Wiley-Liss, 1996), 12. The authors note: "Although life is a continuous process, fertilization is a critical landmark because, under ordinary circumstances, a new, genetically distinct human organism is thereby formed." Ibid., 8.

10. Lee Silver, *Remaking Eden: Cloning and Beyond in a Brave New World* (New York: Avon Books, 1997), 39.

11. M. Alikani et al., "Monozygotic Twinning in the Human Is Associated with the Zona Pellucida Architecture," *Human Reproduction* 9 (1994): 1318–21.

12. R. L. Gardner, "The Early Blastocyst Is Bilaterally Symmetrical and Its Axis of Symmetry Is Aligned with the Animal-Vegetal Axis of the Zygote

in the Mouse," *Development* 124 (1997): 289–301; R. Beddington and E. Robertson, "Axis Development and Early Asymmetry in Mammals," *Cell* 96 (1999): 195–209.

13. K. Piotrowska et al., "Role for Sperm in Spatial Patterning of the Early Mouse Embryo," *Nature* 409 (2001): 517–21; R. L. Gardner, "Specification of Embryonic Axes Begins before Cleavage in Normal Mouse Development," *Development* 128 (2001): 839–47.

14. K. Piotrowska et al., "Blastomeres Arising from the First Cleavage Division Have Distinguishable Fates in Normal Mouse Development," *Development* 128 (2001): 3739–48.

15. H. Pearson, "Your Destiny, From Day One," *Nature* 418 (4 July 2002): 14–5.

16. It is now known that the nucleus of any body cell can be reprogrammed to form the basis for a new embryo by placement into an enucleated egg (somatic cell nuclear transfer cloning). There are now many successes in redirecting adult stem cells to form other cell types. See, for example, D. Josefson, "Adult Stem Cells May Be Redefinable," *British Medical Journal* 318 (30 January 1999): 282. And PPL Therapeutics in Scotland claims it has developed a way to direct any specialized body cell to form a relatively unspecialized stem cell, then redirect that stem cell to form other cell types. "PPL Follows Dolly with Cell Breakthrough," *Financial Times,* 23 February 2001.

17. National Institutes of Health (NIH), *Stem Cells: Scientific Progress and Future Research Directions* (Department of Health and Human Services, June 2001), ES-9.

18. The commission acknowledged that the act of deriving embryonic stem cells destroys the embryo, and it was skeptical of the claim that derivation of the cells and their later use could ethically be "neatly separated." NBAC, *Ethical Issues in Human Stem Cell Research,* 1:49, 54–55, 70–71. In fact, the commission recognized "the close connection in practical and ethical terms between derivation and use of the cells." Ibid., v.

19. "National Institutes of Health Guidelines for Research Using Human Pluripotent Stem Cells," *Federal Register* 65 (25 August 2000): 51976–81, at 51980.

20. Glenn McGee, quoted in J. Spanogle, "Transforming Life," *The Baylor Line* (Winter 2000): 34. Also see Lee Silver, "Watch What You Are Calling an Embryo," in *Washington Post,* 19 August 2001, B4.

21. NBAC, *Ethical Issues in Human Stem Cell Research,* 1:53.

22. See U.S. Code, 42 USC §289g.

23. Says Glenn McGee: "Pretending that the scientists who do stem cell research are in no way complicit in the destruction of embryos is just wrong, a smoke and mirrors game on the part of the NIH. It would be much better to take the issue on directly by making the argument that destroying

embryos in this way is morally justified—is, in effect, a just sacrifice to make." Quoted in J. Spanogle, "Transforming Life," 30.

24. Editorial, "Downside of the Stem Cell Policy," in *New York Times,* 31 August 2001, A18.

25. "The logic of the American free enterprise system suggests that President Bush's decision is going to provide incentive for the private sector to get more involved. And once the basic research is conducted, the private sector likely will have great incentive to step in and transform this basic research into therapies for disease." Testimony of Health and Human Services Secretary Tommy Thompson before the Senate Health, Education, Labor, and Pensions Committee, 5 September 2001 (www.hhs.gov/news/speech/2001/010905.html).

26. NBAC, *Ethical Issues in Human Stem Cell Research,* 1:53.

27. Testimony of Christopher Reeve before the Senate Appropriations Subcommittee on Labor, Health, and Human Services and Education, 26 April 2000.

28. Dale Woodbury et al., "Adult Rat and Human Bone Marrow Stromal Cells Differentiate into Neurons," *Journal of Neuroscience Research* 61 (2000): 364 (emphasis added).

29. Darwin Prockop, "Stem Cell Research Has Only Just Begun" (letter), *Science* 293 (13 July 2001): 211–12 (citations omitted).

30. Sylvia Westphal, "The Ultimate Stem Cell," *New Scientist* (26 January 2002): 4. Also see D. S. Krause et al., "Multi-Organ, Multi-Lineage Engraftment by a Single Bone Marrow-Derived Stem Cell," *Cell* 105 (2001): 369–70.

31. Joseph D'Agnese, "Brothers with Heart," *Discover* (July 2001): 36–43; 102.

32. For documentation, see www.stemcellresearch.org, the website of Do No Harm: The Coalition for Americans for Research Ethics, especially "Current Clinical Applications of Adult Stem Cells" (www.stemcellresearch.org/currentaps.pdf) and "Letter to Ruth Kirschstein, Ph.D., Acting Director of the National Institutes of Health" (www.stemcellresearch.org/kirschstein.pdf).

33. M. Cavazzana-Calvo et al., "Gene Therapy of Human Severe Combined Immunodeficiency (SCID)-X1 Disease," *Science* 288 (28 April 2000): 669–72.

34. K. Foss, "Paraplegic Regains Movement after Cell Procedure," *Globe and Mail,* 15 June 2001, A1.

35. E. Ryan et al., "Glycemic Outcome Post Islet Transplantation," abstract #33–LB, Annual Meeting of the American Diabetes Association, 24 June 2001. See http://207.78.21.41/am01/AnnualMeeting/abstracts/PrintResults.asp?idAbs=33–LB.

36. L. Turka, "2001: Innovations in Transplantation," *Medscape Transplantation* 3 (2001), www.medscape.com/viewarticle/421536_3.

37. T. Kirn, "Cord Stem Cells: Results, Not Hype," *Ob. Gyn. News* 36 (15 September 2001), 1.

38. J. Chen et al., "Intravenous Administration of Human Umbilical Cord Blood Reduces Behavioral Deficits after Stroke in Rats," *Stroke* 32 (2001): 2682–88.

39. NIH, *Stem Cells,* ES-1, 23.

40. Jon S. Odorico et al., "Multilineage Differentiation from Human Embryonic Stem Cell Lines," *Stem Cells* 19 (2001): 193–204.

41. Nadya Lumelsky et al., "Differentiation of Embryonic Stem Cells to Insulin-Secreting Structures Similar to Pancreatic Islets," *Science* 292 (2001): 1389–94. With apparently unintentional humor, the NIH describes the findings as follows: "When the cells were injected into diabetic mice, they survived, although they did not reverse the symptoms of diabetes." The NIH means that the cells survived; the mice all died. NIH, *Stem Cells,* 73–74.

42. Odorico et al., "Multilineage Differentiation from Human Embryonic Stem Cell Lines," 200.

43. NIH, *Stem Cells,* 17. The NIH report notes on page 97: "From the perspective of toxicology, the proliferative potential of undifferentiated human embryonic and embryonic germ cells evokes the greatest level of concern. A characteristic of human embryonic stem cells is their capacity to generate teratomas when transplanted into immunologically incompetent strains of mice. Undifferentiated embryonic stem cells are not considered as suitable for transplantation due to the risk of unregulated growth."

44. E. Jonietz, "Innovation: Sourcing Stem Cells," *Technology Review* (January/February 2001), 32.

45. GretchenVogel, "Can Old Cells Learn New Tricks?" *Science* 287 (2000): 1419; L. Johannes, "Adult Stem Cells Have Advantage Battling Disease," *The Wall Street Journal,* 13 April 1999, B1.

46. NIH, *Stem Cells,* 17; also see 63 (any possible advantages of embryonic cells remain to be determined) and 102 (not known whether these cells are better suited for gene therapy).

47. U.S. Catholic Conference News Release, "New Poll: Americans Oppose Destructive Embryo Research, Support Alternatives," 8 June 2001 (www.usccb.org/comm/archives/2001/01-101.htm).

48. Ronald Green, "Toward a Copernican Revolution in Our Thinking about Life's Beginning and Life's End," *Soundings* 66 (1983): 152–57.

49. Ronald Green, in Transcript of the NIH Human Embryo Research Panel (National Institutes of Health, Rockville, Md., 1994), 11 April 1994, 92.

50. Joseph Fletcher, *Situation Ethics: The New Morality* (Philadelphia: Westminster Press, 1966), 120–21.

51. While some non-Catholics cite Catholic philosopher Norman Ford in defense of the theory of delayed ensoulment, for example, Dr. Ford himself holds that, as different theories continue to be debated, the church is

absolutely right to maintain "that the fruit of human generation . . . is morally inviolable from conception and human embryos should be treated as persons." Norman Ford, "The Human Embryo as Person in Catholic Teaching," *The National Catholic Bioethics Quarterly* 1 (2001): 160.

52. Representative James McDermott, quoted in *Congressional Record* (31 July 2001): H4922.

CHAPTER 10

What Would You Do If . . . ?

Human Embryonic Stem Cell Research and the Defense of the Innocent

M. THERESE LYSAUGHT

Into whatever city you go, after they welcome you, eat what they set before you, and cure the sick there. Say to them, The reign of God is at hand (Luke 10:9).[1]

This passage, and St. Luke's continuing presence to us in the communion of saints, issues an important reminder that should shape our inquiry into the ethics of human embryonic stem cell research. That reminder is this: healing is a sign of the Kingdom of God. Healing was a fundamental component of Jesus' ministry, as witnessed in the gospels. Healing is central to God's identity as disclosed through revelation. As this particular passage from Luke notes, healing is part of the commission Jesus gives to those he sends out into the world to preach the good news of the kingdom. Healing, therefore, ought to be central to the ways of discipleship and Christian reflection today.

The centrality of healing to the mission of Christian discipleship is witnessed not only in Scripture but in the historic commitment of the Roman Catholic tradition to the practice of healing and support of health. Nowhere is this commitment more evident than in the marked presence of Catholic hospitals and allied health care organizations. The origin of hospitals can be traced to Christian practices of caring for the sick, and for centuries communities of religious women and men in the church have dedicated themselves to the apostolate of caring for the sick and the dying.[2] Currently, Catholic hospitals constitute over 16 percent of all community hospital beds and admissions in the United States. Not simply an ideal, the Catholic commitment to healing is concretely embodied and enacted in our contemporary context.[3]

I begin with this reminder because the Christian commitment to healing is often obscured or ignored by those who caricature and dismiss Catholic arguments against human embryonic stem cell research. The arguments of Catholics or other groups who inveigh against human embryonic stem cell research, in the words of Glenn McGee and Arthur Caplan, are illogical and bizarre. McGee and Caplan accuse opponents of holding that embryos are special people who can never be allowed to die and of ascribing to embryos a sort of super status that outweighs the needs of others in the community.[4] Not only do such claims distort the arguments in question, but they abstract Catholic claims and arguments against human embryonic stem cell research from the broader narratives and practices out of which they emerge. This cannot but render them unintelligible. In order to avoid such misrepresentation, we need to be mindful of the centrality of healing to the practice of the Christian life and the historic embodiment of this commitment in the Catholic tradition in the broader context of the debate about the moral propriety of human embryonic stem cell research.

This said, in this paper I will examine what has emerged as the central moral question surrounding human embryonic stem cell research, at least within the public debate.[5] The question has been phrased in different ways, so I will offer three versions. First, Kenneth Woodward summarizes the issue in *Newsweek:*

What value should we place on human embryos, he asks, and how should their well-being be balanced with that of the millions whose acute suffering might be alleviated through stem cell research and development?[6] The logic of this appeal is undilutedly utilitarian. But, as savvy proponents of human embryonic stem cell research know, utilitarian calculus, while inescapably operative for most moral agents, is generally deemed insufficient, especially when human lives occupy both sides of the equation. Consequently, a second appeal is often launched, one that more subtly individualizes the question. It is usually presented as an image or a narrative rather than as a direct question. Those who followed the controversy as it evolved may remember Mollie and Jackie Singer, 12-year-old twins who spoke at a congressional hearing in July 2001, urging President Bush to permit federal funding for human embryonic stem cell research. Mollie is afflicted with diabetes, and Jackie appealed for stem cell research to advance in order that her sister might be spared the debilitating effects of the disease.[7] Or one may remember the photo dominating the extended coverage by *The New York Times* of President Bush's decision the Sunday after his announcement. In the photo, Charles and Jeri Queenan and their four children soberly watch Bush's August 9th address. The Queenans' daughter Jenna, also twelve years old, struggles with juvenile diabetes, too, and they hope human embryonic stem cell research might cure her.[8]

Mollie, Jackie, Jenna—this second appeal comes in the images and stories of children whose acute suffering might be alleviated through stem cell research. The crux of this appeal is simple. The images whisper: What if this were your child? Indeed, this question is not only whispered. Sooner or later, in any effort to question the moral propriety of human embryonic stem cell research, one can expect a challenge that seems, for the challenger, to be the moral trump card: What would you do if one of your children needed therapy generated by human embryonic stem cell research? What if your child had a terrible disease, and stem cell research provided the only or best possible hope for the alleviation or eradication of the disease? Could you stand against it then?[9] The challenge brings argument to an end. Only a moral barbarian could argue against pursuing a therapy that could possibly relieve the suffering or forestall the early death of a child, particularly one's own child.

Prescinding for a moment from the obvious emotive appeal to feelings of parental succor and obligation, one could argue that this challenge, as well as the utilitarian version of the question stated earlier, paints the situation as one of defense of the innocent. Here we have an innocent: a family member, a child, a multitude that is threatened by an aggressor (in this case, a disease).[10] The individual is appealed to as the one who has the power or ability to come to the defense of the innocent victim.[11] The defense of the innocent victim against the aggressor requires, unfortunately, the sacrifice of a human life.[12]

Is this a situation where the sacrifice of human life might be justified? McGee and Caplan, offering a third version of our question, claim that the central moral issues in stem cell research have to do with the criteria for moral sacrifices of human life.[13] What might such criteria look like? Where might we find moral criteria for justifying the sacrifice of one human life in order to save another or to protect the common good?

Three classic examples, centrally located within the Christian tradition, provide a starting point from which to begin to address this question. These are: (1) the justification of self-defense, offered in one instance by Thomas Aquinas; (2) the classic situation of defense of one's family member or neighbor against a malicious attacker, helpfully analyzed by the late Mennonite theologian John Howard Yoder; and (3) the just war tradition.[14]

These three situations share certain structural features with the current debate. First, in each situation, an "innocent" (i.e., the self, the family member, one's nation) has been or is being attacked. Second, in each situation, the taking of human life is presented as the only, primary, or last option, and it is required to defend the life of an "innocent" third party. Thus, each scenario can be described as one in which the taking of human life might be justified in defense of the innocent, and each provides a classic site within the Christian tradition where moral theologians have struggled with the question of the justified taking of human life.

One might object that these analogies will be of limited relevance to human embryonic stem cell research insofar as they concern, not health care, but violence or war. I would suggest, however, that they are fitting for precisely this reason. For the rhetoric surrounding the human embryonic stem cell debate is rife with images of war. This is not, of course, necessarily specific to the human embryonic stem cell debate: much of this sort of rhetoric arises whenever a new biotechnology is developed and needs to be sold to political and public audiences in the U.S. While I will not create an exhaustive account of this here, a few examples will illustrate.

Consider, for example, McGee and Caplan's article, "The Ethics and Politics of Small Sacrifices in Stem Cell Research." One finds at least seven war-related images in as many pages. Those who seek to develop therapies from human embryonic stem cells are characterized as fighting a just war, a war against suffering caused by the whole gamut of diseases from Parkinson's to cancer to heart disease and more.[15] The annual mortality of cancer, which might potentially be alleviated through human embryonic stem cell research, is compared to the number of people killed in both the Kosovo and Vietnam conflicts.[16] Human embryonic stem cell research advocates plan to sacrifice embryos for a revolutionary new kind of research.[17] Parkinson's disease is likened to a dictator dreaming up the most nefarious chemical war campaign.[18] Resonating with our current political situation, they note that adults and even children are sometimes forced to give life, but only in the defense, or at least interest, of the community's highest ideals and most pressing interests.[19]

McGee and Caplan are far from alone in employing this sort of rhetoric to frame the discussion about human embryonic stem cell research. For many, and certainly for the media, clinical medicine through the auspices of biotechnology is engaged in a war against disease, disability, suffering, and death. Regenerative technologies are referred to as revolutionary. The tools of research and the clinic are the medical armamentarium. Those who suffer from particular illnesses are survivors. Moreover, the hyperdrive politicization of this current issue points to the familiar adage that politics is but war waged by other means. As Katharine Seelye notes, on August 9, 2001, when George W. Bush finally revealed his decision about federal funding of human embryonic stem cell research, they chose to have Mr. Bush announce his decision in prime time on national television, a format that presidents traditionally reserve for explaining military actions or trying to extract themselves from difficult political binds.[20]

This rhetoric of war is, I think, not accidental. In a time of war, different rules apply. Rights and lives can be abrogated in ways that would be considered an outrage in peacetime. For reasons that will become clear, I would challenge the metaphor of war as the proper way of framing our understanding of clinical research. Yet that argument must wait. Instead, for the moment I will accept the terms of the debate offered by advocates of human embryonic stem cell research: that we are at war and that this creates a situation in which the sacrifice of human life may, nay must, be justified.

If so, those who earnestly seek to justify the sacrifice of human life on moral grounds and who wish to do so in terms that transcend bald utilitarianism would do well to begin with traditional arguments that justify such sacrifice in analogous contexts. Traditional arguments have stood the test of time, have proved their power by admitting analogous transfer in other contexts, and have done so in a way premised on substantive moral claims. Should human embryonic stem cell research fit with the structure of these arguments, a compelling case could be made to advance its cause. With this in mind, I turn now to consider the three analogies outlined above:

(1) Aquinas' justification of self-defense; (2) the defense of one's family member or neighbor against a malicious attacker; and (3) just war. Each of these cases could be the subject of this paper in its own right, and my remarks will therefore be far from exhaustive. Instead, I will highlight the morally relevant features of each case and show how they illuminate the rhetoric that attends human embryonic stem cell research.

THOMAS AQUINAS AND THE JUSTIFICATION OF SELF-DEFENSE

A first case where the Christian tradition has permitted the sacrifice of one human life to save another is self-defense. The question of self-defense is worth examining not only as an instance where killing might be justified in defense of the innocent (i.e., the self), but insofar as arguments for the natural right to self-defense and protection of the common good form the basis of the just war tradition that will be examined below.

The classic treatment of self-defense is found in Thomas Aquinas's *Summa Theologica* (II–II, q. 64, a. 7). Here Aquinas considers the question: Whether it is lawful to kill a man in self-defense? After noting that the tradition does not speak with one voice to this question, he concludes that it can be not unlawful. He notes:

> Nothing hinders one act from having two effects, only one of which is intended, while the other is beside the intention. Now moral acts take their species according to what is intended, and not according to what is beside intention, since this is accidental as explained above (43, 3; I–II, 12.1). Accordingly, the act of self-defense may have two effects, one is the saving of one's life, the other is the slaying of the aggressor. Therefore, this act, since one's intention is to save one's own life, is not unlawful, seeing that it is natural to everything to keep itself in being, as far as possible. And

> yet, though proceeding from a good intention, an act may be rendered unlawful, if it be out of proportion to the end. Wherefore, if a man, in self-defense, uses more than necessary violence, it will be unlawful: whereas if he repel force with moderation, his defense will be lawful. . . . Nor is it necessary for salvation that a man omit the act of moderate self-defense in order to avoid killing the other man, since one is bound to take more care of one's own life than of another's. But as it is unlawful to take a man's life, except for the public authority acting for the common good as stated above (3), it is not lawful for a man to intend killing a man in self-defense, except for such as have public authority, who while intending to kill a man in self-defense, refer this to the public good, as in the case of a soldier fighting against the foe, and in the minister of the judge struggling with robbers, although even these sin if they be moved by private animosity.[21]

Aquinas's analysis provides two possible starting points for those interested in developing criteria for sacrificing one human life for the sake of another, specifically, intention and public authority.

Intention, for Aquinas, does not in itself justify an act, in this case, the act of self-defense. Rather, intention is that aspect of an action by which we can determine how it ought to be described or categorized. As any good ethicist knows, 90 percent of the solution to a question lies in how it is described or (we could say) narrated. Our descriptions locate questions within a larger narrative, placing the question in proper relationship to relevant substantive claims that, taken together, point to the morally pertinent dimensions of the issue.

In this case, then, an action whose direct intention is to save one's own life is (somewhat tautologically) properly categorized as an act of self-defense. Self-defense is justified by a broader web of concepts within Aquinas's system: the natural propensity toward self-preservation, our duty to care for one's own life more than for another's, the virtue of justice (under which this discussion is located), and so on. Might advocates of human embryonic stem cell research be able to define the intention of the practice such that it naturally falls under a category that finds itself justified in relationship to substantive moral claims present in contemporary culture? Clearly, advocates argue that, while human embryonic stem cell research requires the destruction of embryos, the intention of ameliorating suffering and preserving the lives of those with serious illness ought to locate it under a different heading—for example, promotion of the common good.

Equally interesting, Aquinas allows public authorities to do what an individual cannot do, namely, to intend to kill a man in self-defense. In order for them to do so lawfully, they must refer the action to the public good. Given the recent controversy over the role the federal government ought to play in funding and oversight of human embryonic stem cell research, advocates might make a case that a Thomistic framework could support the claim that human embryonic stem cell research would be more properly administered by public authorities aiming at the common good—i.e., the NIH and federal funding—than by the private sector. However, while the traditional case for self-defense seems to hold promise for constructing a justification for human embryonic stem cell research, the analogy between such research and self-defense breaks down at a significant number of points, rendering the self-defense argument of doubtful utility.

First, the act or practice of human embryonic stem cell research and an act of self-defense are structurally quite dissimilar. Most obviously, human embryonic stem cell research lacks the binary nature of the act of self-defense: it is necessarily mediated by third parties (researchers, lab technicians, physicians). Moreover, for Aquinas, in an act of self-defense the one justifiably killed is an aggressor. Human embryos clearly are not. For Thomas, even public authorities are limited in their ability to sacrifice life for the common good, being granted permission by Aquinas only to take the lives of aggressors and sinners (II–II, q. 64, a. 3).

Second, it is clear that in Aquinas's analogy, the effects of the one act are immediately related, if not simultaneous: in the same action by which I defend myself I simultaneously kill you. It is this simultaneity that allows Thomas to create what would otherwise rightly be called a fiction—the claim that there is only one direct intention, in spite of the two inseparable effects. As the two effects of an act become separated from each other in time, with subsequent actions required to effect the second outcome, our ability to ascribe a single intention disappears. Some might wish to construe human embryonic stem cell research as one act or practice that has two inseparable effects: one desired and intended, the relief of suffering and the avoidance of death, and one not desired and therefore not directly intended—the destruction of embryos. However, given that these two effects are far removed from each other in time, the legitimacy of this move becomes doubtful.

Third, the intention to save one's own life—while helping one place the action in the proper moral category—is not itself sufficient to render the act lawful. As he notes, "and yet, though proceeding from a good intention, an act may be rendered unlawful, if it be out of proportion to the end. Wherefore if a man, in self-defense, uses more than necessary violence, it will be unlawful: whereas if he repel force with moderation his defense will be lawful." Rather than being a loophole through which one might justify violence, Aquinas is clearly concerned not to give license even toward the pursuit of a good end. The violence that is justified must be *necessary* to save one's own life. If, by any means, violence or the death of the aggressor may be avoided, the act becomes unlawful. With regard to human embryonic stem cell research, the *necessity* of using embryonic stem cells and the ready availability of promising alternatives is precisely what is at issue. I will discuss both of these in more detail below.

In the interest of space, I will simply mention, rather than elaborate on, three additional points of difference. For Aquinas, a justified act of self-defense is an exception for both individuals and for public authorities. As Paul Ramsey notes: he does not say that it is intrinsically right to intend to kill an onrushing, unjust assailant, and then apply this general rule to the case of action in defense of the common good. Intending to kill a man as a means to the public good is clearly an exception to the basic rule (which still remains in force) that no Christian shall intend to kill any man.[22] Relatedly, Aquinas is here attempting to justify actions, not practices. As exceptions, these are seen as ad hoc, one time, unavoidable acts—not as a systematically developed program of activity. Likewise, the actions are considered retrospectively rather than prospectively. The question is: Is this action that has already occurred, unfortunate though it may be, justifiable? The requirements of intention, simultaneity, and proportion render it difficult to imagine how one might prospectively structure an act or practice that would not fall short on any of these measures.[23]

Even the promise of intention dissolves upon closer analysis. For Aquinas, once intention shifts from self-defense to any other intention, it becomes immediately unjustified. In the case of human embryonic stem cell research, advocates identify a range of possible uses for stem cell lines (e.g., basic research into the processes of human development, the testing of cosmetics and household products, and so on) in addition to curing diseases and saving lives. Most if not all of these additional outcomes will likely be more immediate. Moreover, as has been the case with so many other recent developments in biotechnology over the past fifteen years, it is more likely than not that we will find ourselves faced with yet another instance of what one might call the therapeutic shift, wherein the initial rhetoric presented in order to marshal public opinion and funding focuses almost exclusively on the therapeutic potential of the new technology in question. After securing public support and becoming feasible, however, the technology takes on a life of its own and becomes made available for any purpose for which those with money can pay.[24]

In the end, the classical justification of self-defense, as found in Aquinas, fails to provide a moral framework for the sacrificing of one human life for the sake of another in the practice of human embryonic stem cell research. Instead, it offers a framework that seeks to minimize the violence we might naturally inflict on one another in the name of our own needs, desires, or even justice.

WHAT WOULD YOU DO IF . . .?

A second case where some within the Christian tradition have attempted to justify sacrificing one life to save another would be that of killing an assailant in order to defend not the self but an innocent third party. This question is often raised, as John Howard Yoder notes, as a rejoinder to pacifist objections to war. As he observes at the beginning of his short book *What Would You Do?*:[25]

> Sooner or later, in almost any serious discussion about peace and war, someone is sure to ask the standard question: "What would you do if a criminal, say, pulled a gun and threatened to kill your wife?" (or daughter or sister or mother, whichever one the challenger decides to use). It's uncanny how many persons see this question as a way to test the consistency of the pacifist's convictions that war is wrong.[26]

Yoder tackles this question from two directions. He first unpacks the assumptions implicit in the question, and then goes on to show how the situation of defense of a loved one differs significantly from the situation of war. The analogy, in other words, breaks down.

The parallel in the questions raised between the situations of war and human embryonic stem cell research is uncanny. And like the attempt to analogize the defense of the innocent third party to the question of war, the attempt to draw this analogy to human embryonic stem cell research likewise breaks down.[27] Therefore, rather than proceeding as I did with the question of self-defense (i.e., outlining the analogy, identifying points of contact, and showing how it breaks down), I will instead follow Yoder's lead and analyze the assumptions and dynamics at work in the rhetorical apparatus employed by advocates of human embryonic stem cell research. Yoder identifies six assumptions that underlie the "what would you do if" question. Four will be explored here: determinism, control, knowledge, and alternatives.

Determinism is a problem that afflicts the rhetoric surrounding almost every new development in biotechnology.[28] Not surprisingly, then, we find it in the human embryonic stem cell debate in spades. On a first level, advocates of human embryonic stem cell research paint a scenario that unfolds mechanically. Something like the claim that "millions of people will suffer and die unless human embryonic stem cell research is pursued" is often made explicitly or by implication. For example, Stanford biologist Irv Weissman has been quoted as saying: "Anyone who would ban research on embryonic stem cells will be responsible for the harm done to real, alive, postnatal, sentient human beings who might be helped by this research. Opponents are sacrificing these people to keep from destroying embryos in fertility-clinic freezers that will be thrown out anyway."[29] Or John Gearhart, one of the two researchers whose work initiated the public debate, notes that banning research on embryonic stem cells could make "a lot of people in the future suffer needlessly and maybe even die."[30] The converse, "if we agree to allow the research, these people will be spared" is implied as well.

The argument is not only deterministic in structure, it is also deterministic in time. In making their pitch, biotech advocates often like to work in factors of five, positing clinical therapies "within five years," or "in a decade." Ron McKay, a stem cell expert at the NIH, was, in November 2000, even more optimistic, promising that "in a few months it will be clear that stem cells will regenerate tissues. In two years, people will routinely be reconstituting liver, regenerating heart, routinely building pancreatic islets, routinely putting cells into brain that get incorporated into normal circuitry. They will routinely be rebuilding all tissues."[31]

Such deterministic claims, of course, ignore important components of the situation. Essentially dismissing the wide range of other research endeavors that have been in process for decades, they ignore the possibility that other interventions might be developed to ameliorate the suffering of those afflicted by particular diseases. In creating the fiction of imminent clinical application, they pretend that the untold millions cited will not, most likely, suffer and die an early death from their conditions, since so much of research bears so little clinical fruit. Witness, for example, the unfulfilled promise of gene "therapy." Moreover, these deterministic claims obscure the troubling practical reality that, should therapeutic applications be developed from human embryonic stem cell research, they will probably not be made available to most of the people who could benefit. The intractable issues of access to health care, social justice, and global inequities will not simply evaporate should human embryonic stem cell research bear fruit.

Yoder's second charge is that the challenge "what would you do if" assumes "if not my omnipotence, at least my substantial control of the situation. It assumes that if I seek to stop the attacker, I can. Now in some cases," he admits, "this may be true, but in many it is by no means certain."[32] This assumption likewise animates biotech rhetoric, of which advocacy of human embryonic stem cell research is but one example. The rhetoric assumes that if we seek to remedy a particular disease, we can. It is only a matter of enough money, time, freedom, collaboration, and scientific ingenuity.

Moreover, in the case of human embryonic stem cell research, this unwarranted optimism posits control not only over one particular disease or condition, which might be more realistic and achievable, but over the entire gamut of morbidity and mortality. It is the ultimate panacea, the cure for everything. An historian of biotechnology might caution that human embryonic stem cell research falls in line as only the most recent Holy Grail, a cousin of practices spanning organ transplantation to gene therapy that have met with limited or minimal success.

This is not to suggest that human embryonic stem cell research might not lead to the development of therapeutic options for specific diseases. It very well may. But, as Yoder reminds us, the classic theory of just war (to skip ahead for a moment) requires that the criterion of "probable success" be met before innocent lives can be taken. In light of the difficulties that well-funded, novel therapeutic paradigms have historically encountered, coupled with the primitive state of embryonic stem cell research, the probability of moving from theory to therapy, at least at this time, cannot be predicted.

In making this point, however, I am getting ahead. Before elaborating on the difficulty of characterizing the therapeutic success of human embryonic stem cell research as probable, we need to consider a third assumption, namely, that of knowledge. As Yoder notes, "The 'what if?' question presupposes, if not omniscience, at least full and reliable information."[33] Likewise, the kinds of claims made in support of human embryonic stem cell research require a level of knowledge that is certainly not at hand and may well never be, even should such research be funded. For example, as those pursuing the promise of gene therapy have discovered, what one can coax human cells to do in the laboratory often proves impossible to convince them to do in the human body. After much effort, researchers have succeeded in preventing human embryonic stem cells from differentiating in culture long enough to establish cell lines. This outcome has been achieved. What is still lacking is knowledge of precisely what mechanism is at work in preventing differentiation; how to direct cells to differentiate into specific tissue types; how to control cell growth (suppress tumorogenesis) once differentiation has been achieved; how to get cultured tissues to properly engraft; and then, the most difficult piece, how to get them to achieve function in vivo.

As with the field of gene therapy, the rhetoric advocating human embryonic stem cell research steamrolls ahead, hyping the promise of application, while the state of the science and the fundamental understandings of how relevant processes work is itself embryonic. Without first conducting more basic research, the promise has a higher probability of being broken than fulfilled. Of course, perhaps such knowledge is not necessary. As Nicholas Wade exults: "the magic of regenerative medicine is that the physician does not have to know everything, only how to create the right conditions for the body's cells to respond to the appropriate signals."[34] In addition, one might counter that, without the sacrifice of a few frozen embryos, we will not be able to conduct basic research and gain the knowledge necessary to better envision and enact the end. The response to this claim leads us to the last of Yoder's assumptions, namely, that of alternatives.[35] As Yoder notes, the question of "what if" is designed to limit the respondents' options to two: yes or no, for or against, all or nothing. To set up the discussion as if there were only two possible kinds of outcomes (millions suffer and die vs. all are saved) or only one route (human embryonic stem cell research) to the desired outcome is to prejudice the argument. The situation has been descriptively constructed so as to predispose to a particular outcome.

The posing of alternatives, of course, has been one strategy of those who oppose human embryonic stem cell research. To advance basic science, many call for further animal research, noting that the trajectory in animal studies from in vitro to in vivo to therapy is far from complete. Others call for work to first be completed, or at least further advanced, with adult stem cells before moving to human embryonic stem cells. But the rhetoric of the debate will not brook alternatives. Adult stem cells are dismissed by researchers as not totipotent and therefore deficient; they are dismissed because (ironically enough) not enough research has been done to assess their promise. In the media, adult stem cell research becomes "a canard,"[36] "crap science," or "baloney."[37] Alternative means to a shared goal will not be taken seriously. As in most wars, there will be no negotiations; there is no middle ground. Thus, ironically, advocates of human embryonic stem cell research become absolutist, while their opponents emerge as those searching for a compromise that will seek to achieve the ends of protecting innocent life and of working to ameliorate the suffering and mortality associated with the human condition.

In the end, the crux of the "what if" question, as well as the case made in favor of human embryonic stem cell research, lies largely not in rational argument but in emotional appeal. As Yoder notes, the question

> appeals to family connections and bonds of love so that it becomes a problem of emotions as well as thought. Instead of discussing what is generally right or wrong, it personalizes the situation by making it an extension of my own self-defense. Especially is this emotional dimension of the question more visible when the discussion centers on one's duty to protect someone else. Often the questioner will heighten this aspect of the argument by saying, "Perhaps as a Christian you do have the right to sacrifice your own welfare to be loving toward an attacker. But do you have the right to sacrifice the welfare of others for whom you are responsible?"[38]

Classically, these questions are taken up in the just war tradition, and so to our third analogy I now turn.

JUS IN BELLO: HUMAN EMBRYONIC STEM CELL RESEARCH AND THE JUST WAR AGAINST DISEASE

A third case where the Christian tradition has justified the sacrifice of human life would be the just war tradition.[39] As noted at the outset, the language of the just war is invoked by McGee and Caplan. They attempt to argue that, in human embryonic stem cell research, the essence of the embryo—that is, its DNA—is not destroyed but actually lives on in the cell lines and potential tissues developed therefrom. What is destroyed, they claim, are simply the "inessential components" of the embryo—its cytoplasm, external wall, and mitochondria. The reductionistic and gnostic character of these claims aside, they conclude: "It is difficult to imagine those who favor just war opposing a war against such suffering given the meager loss of a few cellular components."[40]

How might the just war tradition illuminate our question? In the interests of space, I will limit my observations to three. First, of our three analogies, the just war tradition provides the closest fit with the situation of human embryonic stem cell research. In the model for a just war, a nation—a multitude—has been attacked or has had its interests threatened. The war may entail the loss of innocent life in the defense of the innocent and the common good. Those who answer the call to fight do so from a position of innocence, and it is recognized that in pursuing the aggressor, innocent civilians on both sides might be killed as well as combatants. But, at the same time, an obligation to protect those unjustly attacked and to work for justice on their behalf is invoked.[41]

Furthermore, the context of human embryonic stem cell research mirrors a number of *jus ad bellum* criteria, the conditions that must be met for a war to be legitimately declared. One could make a case that the cause is just—humanity has a right to defend itself against the onslaught of disease. The war must be declared by a competent, public authority—in this case, perhaps the NIH. The intention must be right, namely, the restoration of peace—which a world free of the ravages of disease approximates. Success must be probable. Apart from my earlier skepticism about the probability of moving from the laboratory to clinical applications, one could grant, for the sake of argument, that human embryonic stem cell research has a sufficient prospect of probable success. In light of this, one could argue that the principle of proportionality is likewise met—the good expected by pursuing the research outweighs the damages to be inflicted in the loss of embryonic life.[42]

However, two important criteria remain, both of which are essential for validating a particular war as just. The first is a final *jus ad bellum* condition: that all peaceful alternatives must first be exhausted. This is also known as the condition of last resort. The debate over alternatives—further animal studies, the use of adult stem cells or placental stem cells—has been discussed above. Until it can be definitively established that all nonviolent alternatives have been exhausted, that human embryonic stem cell research truly is a last resort, the analogy to a "just war" will fail. This is a process that will take time.

In addition to the exhaustion of all peaceful alternatives as a crucial condition for going to war, the just war tradition also provides conditions that must be met during combat, the *jus in bello* criteria. For our purposes, the key condition is that of discrimination or noncombatant immunity. The principle of discrimination protects the immunity of noncombatants by restricting direct targeting to combatants, military installations, and factories whose products are directly related to the war effort. As Aquinas notes in his discussion of war: "those who are attacked should be attacked because they deserve it on account of some fault."[43] Noncombatants are not to be targeted. Just warriors realize that, in the course of attacking legitimate targets, innocent noncombatants may be killed. But within the tradition, a most important moral distinction obtains between recognizing that noncombatants may accidentally and tragically be killed and directly targeting those noncombatants.

In the case of human embryonic stem cell research, frozen embryos occupy the place in the analogy of noncombatants. It cannot be argued that the loss of embryonic life is an unintended, indirect, and accidental by-product of the activities of the research. For this is what is at stake, the ending of embryonic life—not, contra McGee and Caplan, simply the loss of embryonic identity. Human embryonic stem cell research directly targets the lives of human embryos—frozen though they may be, slated for disposal though they may be—in order to achieve the ends of the war. Within the just war tradition, this means to a good end would not be licit. It would be total war.

Finally, the just war tradition reflects the commitments of the Christian tradition from which it emerged. As Aquinas notes, "Those who wage war justly aim at peace."[44] The imperfect peace obtainable in this world is considered to be the normative human condition, and war is reluctantly admitted into the realm of possibility in order to restore natural order and harmony. Aquinas's discussion of war is located, in the *Summa,* not under the heading of justice, where one might expect to find it, but rather under the heading of charity. War is properly categorized as sin, a vice, a violation of the virtue of charity, of the friendship between humans and God that is, within the human community, made possible by the incarnation. Cognizant of this, the just war tradition seeks not as much to carve out a space for the legitimacy of war but rather to create parameters that will severely limit it.

WAR AND PEACE

In so limiting the legitimate taking of human life in war or self-defense, the Christian tradition fails to provide moral criteria that would justify directly and intentionally taking innocent human life. By illuminating the operative assumptions of the human embryonic stem cell debate, analysis of the classic situation of defense of the neighbor renders that particular analogy similarly unhelpful. In each of the three analogies, a case might be made for taking the life of the aggressor. But no moral criteria emerge that would justify sacrificing the life of one not party to the conflict, even in order to save the life of another. One is free to sacrifice one's own life—one may find oneself called to be a martyr—but neither an individual nor public authorities may justifiably sacrifice the life of even one innocent person, even for the sake of the common good. Therefore, as long as we hold that human embryos qualify as human life, "sacrificing" them is not an available moral option.

Ken Woodward reminds us that "the words we choose to frame our arguments reveal the moral universe we inhabit," and it is with this thought that I would like to close.[45] McGee and Caplan end their article echoing Woodward's claim. They state:

> The issues here are novel and they are hard, but mostly they require philosophical innovation about what an embryo is and how we are to treat embryonic material in a time of stem cell research [one hears the resonance: "in a time of war"]. Our argument here is that no embryo need be sacrificed, but we must alter the terms and goals of our debate to frame an appropriate moral framework for dealing with embryos.[46]

In other words, McGee and Caplan propose to resolve this particular moral controversy by redefining the terms—what an embryo is, what it means to kill. They propose to create a different story to describe what we are doing. This is a classic tactic in wartime: to dehumanize the other, to craft a narrative that justifies the necessary use of lethal force, and to tell ourselves that we do it in order to protect the community's highest ideals and most pressing interests. They suggest that the way out of the dilemma is to descriptively construct the practice of human embryonic stem cell research so as to predispose to a particular outcome.

I cannot but agree that a necessary step forward toward resolving the debate over human embryonic stem cell research is the narrative task of redescription. I opened this paper with a passage from St. Luke, and that passage points to a fundamentally different narrative frame for the debate about human embryonic stem cells, in particular, and biotech and clinical research more generally. St. Luke reminds us that, for Christians, healing is understood not in relation to war but in relation to peace.[47] Healing, that practice rightly privileged as a central and enduring commitment for Christian identity and communities, is not, within a Christian narrative, an end in itself. Rather, healing is a sign of the "reign of God," a practice rooted in the identity and actions of the God of peace. For Christians, the healing that we pursue must be anchored in the broader context of God's work in the world and our participation therein. If we abstract the commitment to healing from its narrative context, we are left with a formal claim that becomes an end in itself, to which any and all means might be fitted, even the means of killing embryos. In the end, to paraphrase Yoder, I do not know what I would do if one of my children needed the products of human embryonic stem cell research. But I know that what I ought to do should be illuminated by the story of the Trinitarian God, whose story is one of peace, healing, and compassion—the difficult activity of suffering with those who suffer precisely because, want as we might, we cannot eliminate that suffering.[48]

NOTES

1. As a theologian and the first speaker in a three-day conference on new frontiers opened in science and ethics by human embryonic stem cell research (and sponsored by Marquette University, the Archdiocese of Milwaukee, and the Wisconsin Catholic Conference), I thought it seemed particularly fitting to begin this paper with a passage from the day's lectionary readings. Little did I anticipate that October 18, 2001—the day the conference opened—would turn out to be the Feast of St. Luke, Evangelist, who was reputed to be (among other things) a physician. Physicians, accordingly, claim him as one of their patron saints.

2. Charles Curran, "Roman Catholic Medical Ethics," in *Transition and Tradition in Moral Theology* (Notre Dame, Ind.: University of Notre Dame Press, 1979), 175.

3. For these and other statistics on the Catholic presence in U.S. health care, see the website of the Catholic Health Association of the U.S. at: www.chausa.org/aboutcha/chafacts.asp.

4. Glenn McGee and Arthur Caplan, "The Ethics and Politics of Small Sacrifices in Stem Cell Research," *Kennedy Institute of Ethics Journal* 9, no.2 (1999): 157, 151.

5. This essay takes the ordinary moral language of the public debate as its starting point. In preparing for the conference, I informally "surveyed"

friends, colleagues, and students, asking them "what do you think about research with human embryonic stem cells?" I was surprised by how often we ended up at the "what would you do if?" question discussed below. As John Howard Yoder notes in his analogous context, "The way the question is put arises very naively and authentically from ordinary language of lay ethical debate" ("What Would You Do If?" *Journal of Religious Ethics* 2, no. 2 [1974]: 82). The anomaly revealed simply in this anecdotal experience led me to the questions posed below, since, as Yoder further notes, "ethical discourse properly arises out of the deepening self-critique of ordinary argumentation." The ordinary language of public discourse as presented in the media powerfully shapes the opinions of so many, especially on issues of bioethics. Insofar as public debate itself is informed and shaped by "bioethics communicators" like Glenn McGee (self-description at the conference "Stem Cell Research: New Frontiers in Science and Research," Milwaukee, 19 October 2001), it provides an important point of entry for engaging both the rhetorical and philosophical components of the discussion.

6. Kenneth L. Woodward, "A Question of Life or Death," *Newsweek* (9 July 2001): 31.

7. Sheryl Gay Stolberg, "Stem Cell Debate in House Has Two Faces, Both Young," *New York Times*, 18 July 2001, A1.

8. John W. Fountain, "Stem Cell Decision Does Not End the Debate," *New York Times*, 12 August 2001, 1, 26. Interestingly, the Queenans and the Singers are listed as Roman Catholics and "devout Catholics," respectively.

9. Sometimes, of course, the question concerns another member of one's family: spouse, parent, sibling. The appeal to one's children is, of course, the most powerful.

10. For an account of disease as an aggressor in the context of a theological response, see Arthur C. McGill, *Suffering: A Test of Theological Method* (Philadelphia: Westminster Press, 1982).

11. Terrence W. Tilley helpfully argues that much confusion in the Catholic attempt to forge a "consistent ethic of life" stems from equivocation on the term "innocent," especially between and within discussions of abortion and just war. He notes that there is a difference between the innocence of moral agents—those who act—and the innocence of moral *patients*—those upon whom an act is performed. See his "The Principle of Innocents' Immunity," *Horizons* 15, no.1 (1988): 43–63. For the purposes of this essay, I will use it in its traditional undifferentiated sense.

12. Throughout this essay, of course, I will presume that human embryos are one of a class of creatures that come under the heading "human life." That this is now questioned is evidenced by the opening of Ken Woodward's question ("what value do we place on human embryos . . . ?"). Others more explicitly raise the question of whether we should consider thawed embryos "alive" or whether embryos prior to twenty-one days even ought to be identified as "organisms." See David Hersenov, "The Problem

of Potentiality," *Public Affairs Quarterly* 13, no.3 (July 1999): 255–71, or his subsequent piece,"An Argument for Limited Human Cloning," *Public Affairs Quarterly* 14, no.3 (July 2000): 245–58. However, if one presumes that human embryos do not qualify as "human life," the main moral question with regard to human embryonic stem cell research essentially evaporates. One might still explore questions of cow-human chimeras or similar entities created through in vitro techniques, but it would render the moral question of human embryonic stem cell research moot. This is one strategy pursued by advocates of the research.

13. McGee and Caplan, "The Ethics and Politics of Small Sacrifices . . . ," 151.

14. One might also look to three analogous situations within the broad umbrella of health care: triage, human experimentation, and maternal-fetal conflict. Each of these situations wrestles with the possibility that one life might be lost or sacrificed in order to benefit others. How is this situation like or unlike these three other situations? Might they provide insight for understanding when the claims of particular human lives might override the concern for the protection of embryonic life? Answers to these questions await a subsequent essay.

15. McGee and Caplan, "The Ethics and Politics of Small Sacrifices . . . ,"156.

16. Ibid., 154.

17. Ibid., 152.

18. Ibid., 156, 154.

19. Ibid., 153.

20. Katharine Q. Seelye, "Bush Gives His Backing for Limited Research on Existing Stem Cells," *New York Times,* 10 August 2001.

21. Thomas Aquinas, *Summa Theologica,* trans. Fathers of the English Dominican Province, 2d rev. ed. (London: Burns, Oates, and Washburn, 1920–1942). Available at www.newadvent.org/summa/.

22. Paul Ramsey, *War and the Christian Conscience* (Durham, N.C.: Duke University Press, 1961), 40–41.

23. Those familiar with the Catholic tradition will have undoubtedly noticed that I have studiously avoided using the phrase "double effect." Though the classic principle of double effect takes its origins from Aquinas's account of self-defense, the principle as now articulated radically departs from his limited account. Since the sixteenth century, the principle has been articulated as an attempt to provide justifications for killing innocent persons. (See Ramsey, ibid., p. 47. Ramsey cites Joseph T. Mangan, "An Historical Analysis of the Principle of Double Effect," *Theological Studies* 10, no.1 [1949]: 41–61.) Such a shift demonstrates the sorts of problems that can occur when one attempts to lift a "principle" out of its narrative context. As mentioned earlier, the narrative context anchors a question within a web

of substantive moral concepts that are necessary for making the argument. If, for Aquinas, it is not just to intend to kill an *unjust* aggressor in order to save one's own life, how much less so would it be to kill innocent life in order to save one's own? Thomas's discussion of self-defense not only does not help us in creating criteria for justifying the sacrifice of innocent human life; it provides a compelling argument against it.

24. One might less charitably refer to this as "the therapeutic bait-and-switch." Examples of technologies that argue from the therapeutic premise would be gene "therapy" (the promise embedded in the very term), the cloning rhetoric that followed upon Dolly and other ventures in the 1990s, the development of sperm-sorting techniques for sex selection, and so on. Sperm sorting, or "Microsort" as it is marketed, is an example of how quickly a developed technique can leave its "therapeutic" context and be made available for other purposes.

25. John Howard Yoder, *What Would You Do?* (Scottsdale, Pa.: Herald Press, 1983). See also Yoder, "'What Would You Do If . . . ?' An Exercise in Situation Ethics," *Journal of Religious Ethics* 2, no. 2 (1974): 81–105. Gilbert C. Meilaender also draws on Yoder's essay in his testimony before the National Bioethics Advisory Commission (NBAC). See *Ethical Issues in Human Stem Cell Research*, 3 vols. (Rockville, Md.: June 2000), 3:E1–E6.

26. Yoder, *What Would You Do?*, 13.

27. For example, as with the analogy to self-defense, embryos cannot be properly described as "aggressors," which is morally relevant for this second situation. Likewise, the situation of defense of the innocent compels agreement because of the immediacy and magnitude of the harm that will befall the victim. The killing of the aggressor is allowed in order to prevent a harm from occurring, not to redress a harm that has already taken place. And so on.

28. Yoder identifies three deterministic elements of the standard question. First, "the way the question is usually asked assumes that I alone have a decision to make." Second, the scenario "unfolds mechanically"; once the situation is engaged, the actions of the actors are predetermined. Neither the potential attacker nor the potential victim can exercise any other role than the one predetermined. Third, "the assumption is that how I respond solely determines the outcome of the situation." In the end he notes, "This deterministic assumption is in some sense self-fulfilling. If I tell myself there are no choices, there are less likely to be other choices. Still less will I feel a creative capacity (or duty) to make them possible if I don't expect them. But then the limit is in my mind, not in the situation." Yoder, *What Would You Do?*, 14–15.

29. *Newsweek* (9 July 2001): 24. See also McGee and Caplan, "The Ethics and Politics of Small Sacrifices . . . ," 153–54: "Stem cell research consortium Patient's CURe estimates that as many as 128 million Americans

suffer from diseases that might respond to pluripotent stem cell therapies. Even if that is an optimistic number, many clinical researchers and cell biologists hold that stem cell therapies will be critical in treating cancer, heart disease, and degenerative diseases of aging such as Parkinson's disease. More than half of the world's population will suffer at some point in life with one of these three conditions, and more humans die every year form cancer than were killed in both the Kosovo and Vietnam conflicts."

30. *Newsweek* (9 July 2001): 27.

31. Nicholas Wade, *Life Script* (New York: Simon and Schuster, 2001), 121.

32. Yoder, *What Would You Do?*, 15.

33. He continues: "Not only does it assume on my part that events will unfold in an inevitable way, but it also presumes that I am reliably informed about what that unfolding will be like. I know that if I do not kill the aggressor, he will rape my wife, kill my daughter, attack me, or whatever. And I know I will be successful if I try to take his life." Ibid., 16–17.

34. Wade, *Life Script*, 168.

35. I am here collapsing his discussion of "other options" under the heading of "alternatives."

36. Sheryl Gay Stolberg, "A Science in Its Infancy, but with Great Expectations for Its Adolescence," *New York Times*, 20 August 2001, A17.

37. *Newsweek* (9 July 2001): 27.

38. Yoder, *What Would You Do?*, 19–20.

39. Gilbert C. Meilaender examines a different set of "war"-related arguments in relation to human embryonic stem cell research, taking as his interlocutors both the NBAC report and McGee and Caplan. See his "The Point of a Ban: Or, How to Think about Stem Cell Research," *Hastings Center Report* 31, no.1 (2001): 9–16.

40. McGee and Caplan, "The Ethics and Politics of Small Sacrifices . . . ," 156. The claims made here are not only reductionistic—reducing human identity to DNA—but also gnostic and dualistic insofar as our actual concrete embodiment is deemed not an essential part of who we are.

41. In other words, the human embryonic stem cell debate may approximate the question: "Can an otherwise neutral nation intervene in defense of an innocent party that is attacked by some other nation?"

42. Again, I am making this latter claim for the sake of argument.

43. ST II–II, q. 40, a. 1.

44. Ibid., reply to obj. 3.

45. Woodward, "A Question of Life or Death," 31.

46. McGee and Caplan, "The Ethics and Politics of Small Sacrifices . . . ," 157.

47. On a similar note, Mark Kuczewski suggested a similar critique of the tendency to construe science and clinical research as a "war against na-

ture," rather than situating it in the context of the "story of creation" (comment at the conference "Stem Cell Research," Milwaukee, 18 October 2001).

48. I would very much like to thank Nancy Snow for inviting me to participate in what was such a vital and thorough conference. It was an honor to be part of such an esteemed slate of presenters and a privilege to be able to offer my thoughts to the Wisconsin Catholic Bishops Conference and the Archdiocese of Milwaukee. I must also thank my colleagues who read and so helpfully commented on previous drafts of this paper: Michael Barnes, Una Cadegan, Dennis Doyle, James Heft, Brad Kallenberg, Jack McGrath, Sandra Yocum Mize, Maureen A. Tilley, and Terrence W. Tilley. In them, I am richly blessed.

Glossary

Adult Stem Cell—A cell taken from mature tissue that can renew itself but has a limited ability to transform into specialized cell types.

Allele—A series of two or more different genes that occupy the same position on a chromosome.

Antigens—Substances capable of inducing a specific immune response and then reacting with the products of this response (i.e., with antibodies, specifically T-lymphocytes) under appropriate conditions.

Antioxidant—A synthetic or natural substance that acts as an agent to inhibit oxidation. It prevents or delays deterioration of a particular product as a result of oxygen in the air.

Assisted reproductive technology—Fertility treatments that involve a laboratory handling eggs or embryos, such as in vitro fertilization.

Autoimmune—Cells and/or antibodies operating in such a way that an individual's immune system begins to react against its own tissues.

Autologous—Refers to what would normally be found occurring in a particular type of tissue or in a particular part of the body.

Blastocyst—A preimplantation embryo of 30 to 150 cells.

Chromosomes—Nucleic acid–protein structures in the nucleus of a cell and composed chiefly of DNA, the carrier of heredity information. Chromosomes contain genes, working subunits of DNA that carry the genetic code for specific proteins. A normal human body contains 46 chromosomes; a normal human gamete, 23 chromosomes.

Clones—A group of organisms (or cells) derived from a single organism (or cell) following asexual reproduction, all of which have identical genetic structures.

Cloning—The process of the creation of an animal or person that derives its genes from a single other individual.

Collagen—The protein substance that comprises the white fibers of skin, tendon, bone, cartilage, and all other connective tissue.

Colony—A group of identical cells that were derived from one parent cell; a set of clones.

Colony-forming cells—Groups of cells growing on a solid nutrient surface with each group being created from the multiplication of an individual cell.

Cytoplasm—The site of most chemical activities for the cell. It is the substance of a cell, exclusive of the nucleus, consisting of a solution containing organelles and inclusions.

Differentiation—The process by which early unspecified cells acquire the features of specific cells such as heart tissue, liver, or muscle.

DNA—Abbreviation for deoxyribonucleic acid. The material that contains the instuctions for making all the parts of the body.

Embryo—The earliest stage of development from the single cell to implantation in the uterus.

Embryonic germ cell—A cell from an embryo that has the potential to become a wide variety of specialized cell types.

Endothelium—A layer of flat cells that line cavities of the heart, blood, and lymphatic vessels.

Engraftment—When bone marrow is infused during a transplant, engraftment occurs when the patient accepts the bone marrow and begins to produce blood cells.

Filament—A thin threadlike form, either segmented or unsegmented.

Filopodial (Filipodium)—A long, slender rod that protrudes from the cell of certain amoebae for the sake of nourishment and movement.

Gamete—Mature germ cell, which unites with another in sexual reproduction.

Gene—A unit of heredity that is a segment of DNA located on a specific site on a chromosome.

Genotype—The genetic constitution of an organism or cell.

Germ cell—Cell in the body of an organism, specialized for reproductive purposes and able to unite with one of the opposite sex to form a new individual.

Glia (synonymous with neuroglia)—Supporting tissues of the brain that are intermingled with elements of nerve tissues. Unlike neurons, they do not conduct electrical impulses, yet they are thought to play an important metabolic function because of their location.

Growth factor—In animals, polypeptide hormones or biological factors produced by the body to stimulate the growth and development of cells, tissues, and organs.

Hematopoietic—Having the quality of the type of stem cell that gives rise to distinct daughter cells. One is a copy of the stem cell, and the other will develop into mature blood.

Hemoglobinopathy—Disorder or disease caused by abnormalities in hemoglobin molecules in the blood.

Hepatocyte—Epithelial liver cell.

Histocompatibility—A state of similarity between tissues of two organisms that permits graft transplantation.

Hydatidiform moles—A fairly rare mass or tumor that can develop in the uterus early in a pregnancy.

Immunogen—See antigens.

Immunogenicity—Having the state or property of being antigenic, that is, capable of inducing a specific immune response, then reacting with the products of this response.

Inner Cell Mass—A group of cells found in the preimplantation embryo of a mammal, giving rise to the embryo and potentially capable of forming all tissues except the trophoblast.

In vitro—Done outside the body.

In vivo—Done within the living body.

Major Histocompatibility Complex (MHC)—"The set of gene loci specifying major histocompatibility antigens" (On-line Medical Dictionary).

Mesenchyme—Meshwork of embryonic connective tissue originating from the mesoderm.

Mesoderm—Middle of the three germ layers (cell layers that compose the early embryo). It gives rise to multiple systems and to connective tissue.

Mitosis—Normal process of somatic reproduction of cells where the nucleus is modified, resulting in the formation of two cells containing the same chromosome and DNA content as the original cells.

Morphology—The science concerned with studying the structure of animals and plants.

Multipotent—Capable of giving rise to some specialized cells or tissues of an organism.

Nestin—"Large intermediate filament protein found in developing rat brain" (On-line Medical Dictionary).

Neuroepithelium—Outer germ level of the embryo that gives rise to nervous system cells.

Nucleus—The core of a cell that contains the chromosomes.

Oocyte—Immature ovum that is still developing prior to its completion and release.

Osteogenesis imperfecta (OI)—A group of genetic diseases of the bones characterized by frail, brittle bones.

Perikaryon—Cell body encompassing the nucleus of a neuron.

Phenotype—The observable characteristics displayed by an organism as a result of the interaction between its genotype and particular environmental factors.

Pluripotent—Capable of giving rise to most tissues of an organism.

Primitive Streak—A ridge in the early embryo achieved by the migration of cells. It appears on day 15 and gives an axis to the developing embryo.

Progenitor cells—The cells from which the cells of a particular tissue are generated through cell division (parent cell).

Receptor—A molecular structure within a cytoplasm, or on the cell surface, that binds to a specific factor, such as a hormone, antigen, or neurotransmitter.

Refractile—Refers to structures within cells that scatter light, making them appear bright.

Somatic cell—Cell of the body other than egg or sperm.

Somatic cell nuclear transfer—The transfer of a cell nucleus from a somatic cell into an egg from which the nucleus has been removed.

Stem cells—Nonspecialized cells that have the capacity to self-renew and to differentiate into more mature cells.

Stromal Cells—Connective tissue cells of an organ.

T-cell (T lymphocyte)—What precursor cells mature into upon moving into the thymus.

Tissue or cell culture—Growth of tissue in a laboratory dish for experimental research.

Totipotent—Having unlimited capability. Totipotent cells have the capacity to specialize into extraembryonic membranes and tissues, the embryo, and all postembryonic tissues and organs.

Trophoblast—Extraembryonic cell layer that forms around the mammalian blastocyst. It connects the embryo to the wall of the uterus and is the source through which the embryo is nourished by the mother.

Type I Collagen—The most abundant variety of the protein substance that composes the fibers of skin, tendons, bones, cartilage, and all other connective tissue. They form large, strong, well-organized strands.

Unipotent—Refers to a cell that can only develop in a specific way to produce a certain end result.

Zygote—A cell formed by the union of male and female germ cells (sperm and egg, respectively).

SOURCES FOR GLOSSARY

National Institutes of Health. *Stem Cells: Scientific Progress and Future Research Directions.* June 2001.

National Institutes of Health. "Stem Cells: A Primer." www.nih.gov/news/stemcell/primer.htm. May 2000.

On-Line Medical Dictionary. http://cancerweb.ncl.ac.uk/omd.

Regalado, Antonio, et al. "Stem-Cell Issue Entangles Science and Policy." *The Wall Street Journal.* 10 August 2001.

Singleton, Paul, and Diana Saisbury, eds. *Dictionary of Microbiology and Molecular Biology.* 3rd ed. New York: John Wiley & Sons, Ltd. 2001.

Stedman's Medical Dictionary. 27th ed. Philadelphia: Lipincott, Williams & Wilkins, 2000.

Stedman's Medical Dictionary. 26th ed. Baltimore, Md.: Williams & Wilkins, 1995.

Sykes, J.B., ed. *The Concise Oxford Dictionary.* 7th ed. Oxford: Clarendon Press, 1982.

Tirri, Rauno, et al. *Elsevier's Dictionary of Biology.* New York: University of Turkhu, 1998.

Select Bibliography

Agnese, J. D. "Brothers with Heart." *Discover* 102 (July 2000).
Aird, W., et al. "Long-Term Cryopreservation of Human Stem Cells." *Bone Marrow Transplantation* 9 (1992).
American Association for the Advancement of Science and the Institute for Civil Society Report. "Stem Cell Research and Applications: Monitoring the Frontiers of Biomedical Research." www.aaas.org/spp/dspp/sfrl/projects/stem/report.pdf.
American Fertility Society. "Ethical Consideration in the Use of New Reproductive Technologies." *Fertility Sterilization* 41 (1986).
Andrews, Lori. "State Regulation of Embryo Research." In *Papers Commissioned for the Human Embryo Research Panel.* National Institutes of Health. Vol. 2. 1994.
Annas, G. A., A. Caplan, and S. Elias. "The Politics of Human Embryo Research—Avoiding Ethical Gridlock." *New England Journal of Medicine* 334 (1996).
Aquinas, Thomas. *Summa Theologica.* Translated by the Fathers of the English Dominican Province. New York: Benziger Brothers, 1948.
Ashley, Benedict M., O.P., and Kevin D. O' Rourke, O.P. *Health Care Ethics: A Theological Analysis.* St. Louis, Mo.: The Catholic Health Association, 1978.
Attarian, H., et al. "Long-Term Cryopreservation of Bone Marrow for Autologous Storage." *Bone Marrow Transplantation* 17 (1996).
Baker, Robert. "Stem Cell Rhetoric and the Pragmatics of Naming." *American Journal of Bioethics* 2 (2002).
Barth, Karl. *Church Dogmatics.* Vol. 3: *The Doctrine of Creation.* Edinburgh: T. & T. Clark, 1961.
Beatty, P. G., et al. "Marrow Transplantaion from HLA-Matched Unrelated Donors for Treatment of Hematologic Malignancies." *Transplantation* 51 (1991).

Beddington, R., and E. Robertson. "Axis Development and Early Asymmetry in Mammals." *Cell* 96 (1999).

Ben-Ozer, S., and M. Vermesh. "Full Term Delivery following Cryopreservation of Human Embryos for 7.5 Years." *Human Reproduction* 14 (June 1999).

Berger, Robert L. "Nazi Science—The Dachau Hypothermia Experiments." In *Medicine, Ethics, and the Third Reich: Historical and Contemporary Issues,* edited by John Michalczyk. Kansas City, Mo.: Sheed and Ward, 1994.

Bjornson, C. R., et al. "Turning Brain into Blood: A Hematopoietic Fate Adopted by Adult Neural Stem Cells *In Vivo*." *Science* 283 (1999).

Blau, H. M., et al. "The Evolving Concept of a Stem Cell: Entity or Function?" *Cell* 105 (2001).

Brazelton, T. R., et al. "From Marrow to Brain: Expression of Neuronal Phenotypes in Adult Mice." *Science* 290 (2000).

Broxmeyer, H. E., et al. "Human Umbilical Cord Blood as Potential Source of Transplantable Hematopoietic Stem/Progenitor Cells." *Proceedings of the National Academy of Sciences* 86 (1989).

Bruder, S. P., et al. "Mesenchymal Stem Cells in Osteobiology and Applied Bone Regeneration." *Clinical Orthopaedics and Related Research* 355S (1998).

Burt, R. K., and A. E. Traynor. "Hematopoietic Stem Cell Transplantation: A New Therapy for Autoimmune Disease." *Stem Cells* 17 (1999).

Cahill, Lisa Sowle. "The Embryo and the Fetus: New Moral Contexts." *Theological Studies* 54 (1993).

Callahan, Daniel, and Cynthia B. Cohen. "Letters: Human Embryo Research: Respecting What We Destroy?" *Hastings Center Report* 31 (2001).

Cameron, H. A., et al. "Differentiation of Newly Born Neurons and Glia in the Dentate Gyrus of the Adult Rat." *Neuroscience* 56 (1987).

Caplan, A. I. "Mesenchymal Stem Cell." *Journal of Orthopaedic Research* 9 (1991).

Carlson, B. *Patten's Foundations of Embryology.* 6th ed. New York: McGraw-Hill, 1996.

atholic Health Association of the U.S.: www.chausa.org/aboutcha/ facts.asp.

Timothy. "Regulating the Commercialization of Human an We Address the Big Concerns?" In *Genetic Informa-*

tion, edited by Ruth F. Chadwick and Alison K. Thompson. New York: Kluwer Academic/Plenum Publishing, 1999.

Cavazzana-Calvo, M., et al. "Gene Therapy of Human Severe Combined Immunodeficiency (SCID)-X1 Disease." *Science* 288 (28 April 2000).

Chen, J., et al. "Intravenous Administration of Human Umbilical Cord Blood Reduces Behavioral Deficits after Stroke in Rats." *Stroke* 32 (2001).

Clark, D. L., et al. "Generalized Potential of Adult, Neural Stem Cells." *Science* 288 (2000).

Cohen, C. B., ed. *New Ways of Making Babies: The Case of Egg Donation.* Bloomington: Indiana University Press, 1996.

Congregation for the Doctrine of Faith. *Instruction on Respect for Human Life in Its Origins and on the Dignity of Procreation: Replies to Certain Questions of the Day.* Boston: Pauline Press, 1987.

Curran, Charles. *The Catholic Moral Tradition Today: A Synthesis.* Washington, D.C.: Georgetown University Press, 1999.

Curran, Charles. "Roman Catholic Medical Ethics." *Transition and Tradition in Moral Theology.* Notre Dame, Ind.: University of Notre Dame Press, 1979.

Curran, Charles, and Richard A. McCormick, S.J., eds. *Moral Norms and Catholic Tradition.* Readings in Moral Theology, no. 1. New York: Paulist Press, 1979.

Davis, Henry, S.J. *Moral and Pastoral Theology.* Vol. 1: *Human Acts, Law, Sin, Virtue.* New York: Sheed and Ward, 1946.

Department of Health. Great Britain. *Stem Cell Research: Medical Progress with Responsibility.* London: June 2000.

Donceel, Joseph, S.J. "Immediate Animation and Delayed Hominization." *Theological Studies* 31 (1970).

Dunkel, I. J. "High-Dose Chemotherapy with Autologous Stem Cell Rescue for Malignant Brain Tumors." *Cancer Investigation* 18 (2000).

Dworkin, Ronald M. *Life's Dominion: An Argument about Abortion, Euthanasia, and Individual Freedom.* New York: Knopf, 1993.

Ende, M., and N. Ende. "Hematopoietic Transplantation by Means of Fetal (Cord) Blood: A New Method." *Virginia Medical Monthly* 99 (1972).

Feinberg, Joel. *Harm to Others: The Moral Limits of the Criminal Law.* New York: Oxford University Press, 1986.

Ferrari, G., et al. "Muscle Regeneration by Bone Marrow-Derived Myogenic Progenitors." *Science* 279 (1998).
Flax, J. D., et al. "Engraftable Human Neural Stem Cells Respond to Developmental Cues, Replace Neurons, and Express Foreign Genes." *Nature Biotechnology* 16 (1998).
Fletcher, Joseph. *Situation Ethics: The New Morality.* Philadelphia: Westminster Press, 1966.
Ford, Norman. "The Human Embryo as Person in Catholic Teaching." *The National Catholic Bioethics Quarterly* 1 (2001).
Freed, C. R., et al. "Transplantation of Embryonic Dopamine Neurons for Severe Parkinson's Disease." *New England Journal of Medicine* 344 (2001).
Furton, Edward J. "The Nebraska Fetal Tissue Case." *Ethics & Medicine* 26 (2001).
Furton, Edward J. "Vaccines Originating in Abortion." *Ethics & Medicine* 24 (1999).
Gage, F. H. "Neurogenesis in the Adult Brain." *Journal of Neuroscience* 22 (2002).
Gage, F. H., P. W. Coates, and T. D. Palmer. "Survival and Differentiation of Adult Neuronal Progenitor Cells Transplanted to the Adult Brain." *Proceedings of the National Academy of Sciences* 92 (1995).
Gage, F. H., J. Ray, and L. J. Fisher. "Isolation, Characterization, and Use of Stem Cells from the CNS." *Annual Review of Neouroscience* 18 (1995).
Gardner, R. "The Early Blastocyst Is Bilaterally Symmetrical and Its Axis of Symmetry Is Aligned with the Animal-Vegetal Axis of the Zygote in the Mouse." *Development* 124 (1997).
Gatti, R. A., et al. "Immunological Reconstitution of Sex-Linked Lymphopenic Immunological Deficiency." *Lancet* 2 (1968).
Gawejski, J. L., et al. "Bone Marrow Transplantation Using Unrelated Donors for Patients with Advanced Leukemia or Bone Marrow Failure." *Transplantation* 50 (1990).
Geron Ethics Advisory Board. "Research with Human Embryonic Stem Cells: Ethical Considerations." *Hastings Center Report* 29 (1999).
Gluckman, E., et al. "Hematopoietic Reconstitution in a Patient with Fanconi's Anemia by Means of Umbilical Cord Blood from an HLA-Identical Sibling." *New England Journal of Medicine* 321 (1989).

Gluckman, E., et al. "Outcome of Cord Blood Transplantation from Related and Unrelated Donors." *New England Jounal of Medicine* 337 (1997).

Gould, E., et al. "Learning Enhances Adult Neurogenesis in the Hippocampal Formation." *Nature Neuroscience* 2 (1999).

Gould, E., et al. "Proliferation of Granule Cell Precursors in the Dentate Gyrus of Adult Monkeys Is Diminished by Stress." *Proceedings of the National Academy of Sciences* 95 (1998).

Green, Ronald. *The Human Embryo Research Debates: Bioethics in the Vortex of Controversy.* New York: Oxford University Press, 2001.

Green, Ronald. "Toward a Copernican Revolution in Our Thinking about Life's Beginning and Life's End." *Soundings* 66 (1983).

Grobstein, C. "The Early Development of Human Embryos." *Journal of Medicine and Philosophy* 10 (1985).

Hay, E. *Regeneration.* New York: Holt, Rinehart and Winston, 1966.

Hayflick, L. "The Limited *In Vitro* Lifetime of Human Diploid Cell Strains." *Experimental Cell Research* 37 (1965).

Hayflick, L., and P. Moorhead. "Serial Cultivation of Human Diploid Cell Strains." *Experimental Cell Research* 25 (1961).

Hersenov, David. "An Argument for Limited Human Cloning." *Public Affairs Quarterly* 14 (2000).

Hersenov, David. "The Problem of Potentiality." *Public Affairs Quarterly* 13 (1999).

Holland, Suzanne, Karen Lebacqz, and Laurie Zoloth, eds. *The Human Embryonic Stem Cell Debate: Science, Ethics, and Public Policy.* Cambridge, Mass.: MIT Press, 2001.

Hollenbach, S.J., "Common Good." In *The New Dictionary of Catholic Social Thought,* edited by Judith A. Dwyer. Collegeville, Minn.: Liturgical Press, 1994.

Horwitz, E.M., et al. "Isolated Allogeneic Bone Marrow-Derived Mesenchymal Cells Engraft and Stimulate Growth in Children with Osteogenesis Imperfecta: Implications for Cell Therapy of Bone." *Proceedings of the National Academy of Sciences* 99 (2002).

Horwitz, E.M., et al. "Transplantability and Therapeutic Effects of Bone Marrow-Derived Mesenchymal Cells in Children with Osteogenesis Imperfecta." *Nature Medicine* 5 (1999).

Howard, M.R., et al. "Unrelated Donor Marrow Transplantation between 1977 and 1987 at Four Centers in the United Kingdom." *Transplantation* 49 (1990).

Humphreys, D., et al. "Epigenetic Instability in ES Cells and Cloned Mice." *Science* 293 (2001).

"Individual and Community Risks of Measles and Pertussis Associated with Personal Exemptions to Immunization." *Journal of the American Medical Association* 284 (December 2000).

Jackson, K. A., et al. "Stem Cells: A Minireview." *Journal of Cellular Biochemistry* S38 (2002).

Jacobs, J.P.; C.M. Jones; and J.P. Baille. "Characteristics of a Human Diploid Cell Designated MRC-5." *Nature* 227 (1970).

Johansson, C.B., et al. "Identification of a Neural Stem Cell in the Adult Mammalian Central Nervous System." *Cell* 96 (1999).

John Paul II. "Address to the 18th International Congress of the Transplantation Society." www.vatican.va/holy_father/john_pdf_jp-ii_spe_20000829_transplants_en.html.

John Paul II. *The Gospel of Life*. Boston: Pauline Books, 1995.

John Paul II. "Remarks to President Bush on Stem Cell Research." *National Catholic Bioethics Quarterly* 1 (2001).

Jonietz, E. "Innovation: Sourcing Stem Cells." *Technology Review* (2001).

Josefson, D. "Adult Stem Cells May Be Redefinable." *British Medical Journal* 318 (1999).

Kelly, Gerald, S.J. *Medico-Moral Problems*. St. Louis, Mo.: Catholic Hospital Association, 1957.

Krause, D., et al. "Multi-Organ, Multi-Lineage Engraftment by a Single Bone Marrow-Derived Stem Cell." *Cell* 105 (2001).

Lebacqz, Karen. "The Elusive Nature of Respect." In *The Human Embryonic Stem Cell Debate*, edited by Suzanne Holland, Karen Lebacqz, and Laurie Zoloth. Cambridge, Mass.: MIT Press, 2001.

Lumelsky, N., et al. "Differention of Embryonic Stem Cells to Insulin-Secreting Structures Similar to Pancreatic Islets." *Science* 292 (2001).

Lundberg, C., et al. "Survival, Integration, and Differentiation of Neural Stem Cell Lines after Transplantation to the Adult Striatum." *Experimental Neurology* 145 (1997).

Lustig, Andrew B. "The Common Good in a Secular Society: The Relevance of a Roman Catholic Notion to the Healthcare Allocation Debate." *Journal of Medicine and Philosophy* 18 (1993).

Maher, Daniel P. "Vaccines, Abortion, and Moral Coherence." *The National Catholic Bioethics Quarterly* 2 (2002).

Mahoney, John. *The Making of Moral Theology: A Study of the Roman Catholic Tradition.* Oxford: Clarendon Press, 1987.

McCormick, Richard A., S.J. "Notes on Moral Theology: 1978." *Theological Studies* 40 (1979).

McGee, Glenn, and Arthur Caplan. "The Ethics and Politics of Small Sacrifices in Stem Cell Research." *Kennedy Institute of Ethics Journal* 9 (1999).

McGill, C. *Suffering: A Test of Theological Method.* Philadelphia: Westminster Press, 1982.

McKay, R. D. "Brain Stem Cells Change Their Identity." *Nature Medicine* 5 (1999).

Meilaender, Gilbert C. "The Point of a Ban. Or, How to Think about Stem Cell Research." *Hastings Center Report* 31 (2001).

Menasché, P., et al. "Myoblast Transplantation for Cardiac Repair." *Lancet* 357 (2001).

Meyer, Michael J., and Lawrence J. Nelson. "Respecting What We Destroy: Reflections on Human Embryo Research." *Hastings Center Report* 31 (2001).

Mezey, E., et al. "Turning Blood into Brain: Cells Bearing Neuronal Antigens Generated *In Vivo* from Bone Marrow." *Science* 290 (2000).

Michjada, M. "Fetal Tissue Tranplantation: Miscarriages and Tissue Banks." In *The Fetal Tissue Issue: Medical and Ethical Aspects,* edited by Peter J. Cataldo and Albert S. Moraczewski, O.P. Braintree, Mass.: The Pope John XXIII Medical-Ethics and Education Center, 1994.

Moore, K., and T. Persaud. *The Developing Human.* 6th ed. Philadelphia: W. B. Saunders, 1998.

National Academy of Sciences (NAS). *Scientific and Medical Aspects of Human Reproductive Cloning.* Washington, D.C.: National Academy Press, 2002.

National Bioethics Advisory Commission. *Ethical Issues in Human Stem Cell Research.* Vols. 1–3. Rockville, Md.: National Bioethics Advisory Commission, 1999–2000.

National Institutes of Health. "Guidelines for Research Using Human Pluriopotent Stem Cells." *Federal Register* (2001).

National Institutes of Health. *Report of the Human Embryo Research Panel.* Bethesda, Md.: National Institutes of Health, 1994.

National Institutes of Health. *Stem Cells: Scientific Progress and Future Research Directions.* Bethesda, Md.: National Institutes of Health, 2001.

National Institutes of Health Revitalization Act of 1993. Pub. L. No. 10–43. Sect 111.107 Stat. 129 (codified at 42 U.S.C. Sec. 498A).

National Research Council Committee on Stem Cell Research. "Stem Cells and the Future of Regenerative Medicine." www.nap.edu/books/0309076307/html.

Noonan, John T., Jr. *Contraception: A History of Its Treatment by the Catholic Theologians and Canonists.* Cambridge, Mass.: Harvard University Press, 1986.

O'Donnell, Thomas J. *Medicine and Christian Morality.* New York: Alba House, 1976.

Odorico, J., et al. "Multilineage Differentiation from Human Embryonic Stem Cell Lines." *Stem Cells* 19 (2001).

Pittenger, M.F., et al. "Multilineage Potential of Adult Human Mesenchymal Stem Cells." *Science* 284 (1999).

Pius XII. "An Address of Pope Pius XII to an International Congress of Anesthesiologists." In *Conserving Human Life.* Braintree, Mass.: The Pope John XXIII Center, 1989.

Quaini, F., et al. "Chimerism of the Transplanted Heart." *New England Journal of Medicine* 346 (2002).

Radin, Margaret Jane. *Contested Commodities.* Cambridge, Mass.: Harvard University Press, 1996.

Ramsey, Paul. *War and the Christian Conscience.* Durham, N.C.: Duke University Press, 1961.

Robertson, John A. *Children of Choice: Freedom and the New Reproductive Technologies.* Princeton, N.J.: Princeton University Press, 1994.

Robertson, John A. "Human Embryonic Stem Cell Research: Ethical and Legal Issues." *Nature Reviews Genetics* 2 (2001).

Robertson, John A. "Symbolic Issues in Embryo Research." *Hastings Center Report* 25 (1995).

Rosen, F.S. "Successful Gene Therapy for Severe Combined Immunodeficiency." *New England Journal of Medicine* 16 (2002).

Rubinstein, P., et al. "Outcomes among 562 Recipients of Placental Blood Transplants from Unrelated Donors." *New England Journal of Medicine* 338 (1998).

Rubinstein, P., et al. "Processing and Cryopreservation of Placental/Umbilical Cord Blood for Unrelated Bone Marrow Reconstitution." *Proceedings of the National Academy of Sciences* 92 (1995).

Sanchez-Ramos, B. A., et al. "Adult Bone Marrow Stromal Cells Differentiate into Neural Cells *In Vitro*." *Experimental Neurology* 164 (2000).

Shamblott, M. J., et al. "Derivation of Pluripotent Stem Cells from Cultured Human Primordial Germ Cell." *Proceedings of the National Academy of Sciences* 95 (1998).

Shannon, Thomas A. "Human Embryonic Stem Cell Therapy." *Theological Studies* 62 (2001).

Shannon, T., and A. Wolter. "Reflections on the Moral Status of the Pre-Embryo." *Theological Studies* 51 (1990).

Silver, L. *Remaking Eden: Cloning and Beyond in a Brave New World.* New York: Avon Books, 1997.

Sloan, Philip R., ed. *Controlling Our Destinies: The Human Genome Project from Historical, Philosophical, Social, and Ethical Perspectives.* Notre Dame, Ind.: University of Notre Dame Press, 2000.

Smith, Russell E. "Formal and Material Cooperation." *Ethics and Medicine* 20 (1995).

Solter, D. "Mammalian Cloning: Advances and Limitations." *Nature Reviews Genetics* 1 (2000).

Somia, N., and I. M. Verna. "Gene Therapy: Trials and Tribulations." *Nature Reviews Genetics* 2 (2000).

Spike, Jeffrey. "Bush and Stem Cell Research: An Ethically Confused Policy." *American Journal of Bioethics* 2 (2002).

Sugarman, J., et al. "Ethical Issues in Umbilical Cord Blood Banking." *Journal of the American Medical Association* 278 (1996).

Tauer, Carol A. "Probabilism and the Moral Status of the Early Embryo." In *Abortion and Catholicism: The American Debate,* edited by Patricia Beattie Jung and Thomas A. Shannon. New York: Crossroad, 1988.

Teichler, Zallen D. "US Gene Therapy in Crisis." *Trends in Genetics* 6 (2000).

Thomason, J. A., et al. "Embryonic Stem Cell Lines Derived from Human Blastocysts." *Science* 282 (1998).

Tilley, Terrence W. "The Principle of Innocents' Immunity." *Horizons* 15 (1988).

Tooley, Michael. "Abortion and Infanticide." *Philosophy and Public Affairs* 2 (1972).

Wade, Nicholas. *Life Script.* New York: Simon and Schuster, 2001.

Westoff, C.F. "Fertility in the United States." *Science* 234 (1986).

Westphal, S. "The Ultimate Stem Cell." *New Scientist* (2002).

Wildes, Kevin Wm., S.J. "The Stem Cell Report." *America* (1999).

World Medical Association. "Declaration of Helsinki." *Code of Federal Regulations* 45 (1983).

Yoder, John Howard. *What Would You Do?* Scottsdale, Pa.: Herald Press, 1983.

Yoder, John Howard. "What Would You Do If? An Exercise in Situation Ethics." *Journal of Religious Education* 2 (1974).

Contributors

IRA B. BLACK, M.D., is Professor and Chairman of the Department of Neuroscience and Cell Biology, University of Medicine and Dentistry of New Jersey—Robert Wood Johnson Medical School.

LISA SOWLE CAHILL, PH.D., is J. Donald Monan Professor of Theology at Boston College.

RICHARD M. DOERFLINGER is Deputy Director of the Secretariat for Pro-Life Activities, U.S. Conference of Catholic Bishops, and Adjunct Fellow in Bioethics and Public Policy at the National Catholic Bioethics Center.

KEVIN T. FITZGERALD, S.J., PH.D., is Dr. David Lauler Chair in Catholic Health Care Ethics, Center for Clinical Bioethics, and Associate Professor, Department of Oncology at Georgetown University Medical Center.

EDWARD J. FURTON, PH.D., is an ethicist at The National Catholic Bioethics Center and editor in chief of the *National Catholic Bioethics Quarterly.*

RONALD M. KLINE, M.D., is a physician and partner in the Children's Center for Cancer and Blood Diseases, Las Vegas, Nevada.

JOHN LANGAN, S.J., PH.D., is Joseph Cardinal Bernardin Professor of Catholic Social Thought at the Kennedy Institute of Ethics, Georgetown University.

KAREN LEBACQZ, PH.D., is Robert Gordon Sproul Professor of Theological Ethics at the Pacific School of Religion/Graduate Theological Union, Berkeley.

M. THERESE LYSAUGHT, PH.D., is Associate Professor in the Department of Religious Studies at the University of Dayton.

DAVID A. PRENTICE, PH.D., is Professor of Life Sciences, Indiana State University, and Adjunct Professor, Medical and Molecular Genetics, Indiana University School of Medicine.

DALE WOODBURY is Assistant Professor in the Department of Neuroscience and Cell Biology, University of Medicine and Dentistry of New Jersey—Robert Wood Johnson Medical School.

Index

NANCY E. SNOW is associate professor of philosophy and Director of Core Curriculum at Marquette University.